I0042052

Differential and Integral Calculus

Theory and Cases

Authored by

Carlos Polanco

Department of Mathematics, Faculty of Sciences
Universidad Nacional Autónoma de México, México

&

Department of Electromechanical Instrumentation
Instituto Nacional de Cardiología "Ignacio Chávez"
México

Differential and Integral Calculus - Theory and Cases

Author: Carlos Polanco

ISBN (Online): 978-981-14-6512-3

ISBN (Print): 978-981-14-6510-9

© 2020, Bentham Books imprint.

Published by Bentham Science Publishers Pte. Ltd. Singapore. All Rights Reserved.

First published in 2020.

BENTHAM SCIENCE PUBLISHERS LTD.
End User License Agreement (for non-institutional, personal use)

This is an agreement between you and Bentham Science Publishers Ltd. Please read this License Agreement carefully before using the ebook/echapter/ejournal (**"Work"**). Your use of the Work constitutes your agreement to the terms and conditions set forth in this License Agreement. If you do not agree to these terms and conditions then you should not use the Work.

Bentham Science Publishers agrees to grant you a non-exclusive, non-transferable limited license to use the Work subject to and in accordance with the following terms and conditions. This License Agreement is for non-library, personal use only. For a library / institutional / multi user license in respect of the Work, please contact: permission@benthamscience.net.

Usage Rules:

1. All rights reserved: The Work is the subject of copyright and Bentham Science Publishers either owns the Work (and the copyright in it) or is licensed to distribute the Work. You shall not copy, reproduce, modify, remove, delete, augment, add to, publish, transmit, sell, resell, create derivative works from, or in any way exploit the Work or make the Work available for others to do any of the same, in any form or by any means, in whole or in part, in each case without the prior written permission of Bentham Science Publishers, unless stated otherwise in this License Agreement.
2. You may download a copy of the Work on one occasion to one personal computer (including tablet, laptop, desktop, or other such devices). You may make one back-up copy of the Work to avoid losing it.
3. The unauthorised use or distribution of copyrighted or other proprietary content is illegal and could subject you to liability for substantial money damages. You will be liable for any damage resulting from your misuse of the Work or any violation of this License Agreement, including any infringement by you of copyrights or proprietary rights.

Disclaimer:

Bentham Science Publishers does not guarantee that the information in the Work is error-free, or warrant that it will meet your requirements or that access to the Work will be uninterrupted or error-free. The Work is provided "as is" without warranty of any kind, either express or implied or statutory, including, without limitation, implied warranties of merchantability and fitness for a particular purpose. The entire risk as to the results and performance of the Work is assumed by you. No responsibility is assumed by Bentham Science Publishers, its staff, editors and/or authors for any injury and/or damage to persons or property as a matter of products liability, negligence or otherwise, or from any use or operation of any methods, products instruction, advertisements or ideas contained in the Work.

Limitation of Liability:

In no event will Bentham Science Publishers, its staff, editors and/or authors, be liable for any damages, including, without limitation, special, incidental and/or consequential damages and/or damages for lost data and/or profits arising out of (whether directly or indirectly) the use or inability to use the Work. The entire liability of Bentham Science Publishers shall be limited to the amount actually paid by you for the Work.

General:

1. Any dispute or claim arising out of or in connection with this License Agreement or the Work (including non-contractual disputes or claims) will be governed by and construed in accordance with the laws of Singapore. Each party agrees that the courts of the state of Singapore shall have exclusive jurisdiction to settle any dispute or claim arising out of or in connection with this License Agreement or the Work (including non-contractual disputes or claims).
2. Your rights under this License Agreement will automatically terminate without notice and without the

need for a court order if at any point you breach any terms of this License Agreement. In no event will any delay or failure by Bentham Science Publishers in enforcing your compliance with this License Agreement constitute a waiver of any of its rights.

3. You acknowledge that you have read this License Agreement, and agree to be bound by its terms and conditions. To the extent that any other terms and conditions presented on any website of Bentham Science Publishers conflict with, or are inconsistent with, the terms and conditions set out in this License Agreement, you acknowledge that the terms and conditions set out in this License Agreement shall prevail.

Bentham Science Publishers Pte. Ltd.
80 Robinson Road #02-00
Singapore 068898
Singapore
Email: subscriptions@benthamscience.net

BENTHAM SCIENCE

CONTENTS

Mathematics has beauty and romance. It's not a boring place to be, the mathematical world. It's an extraordinary place; it's worth spending time there.

– Marcus du Sautoyn
1965

FOREWORD I

Differential Calculus by Carlos Polanco is a clear and nicely written introductory textbook for first year calculus, presenting a valuable reference tool for students as well as for their teachers. The book is very well structured and provides a thorough insight into the significance of differential calculus and its outreach for many applications and disciplines. Especially, the presentation of various examples and case studies will help the readers to deepen their acquired knowledge and to relate the theory to practice. In essence, Carlos Polanco's Differential Calculus is a stimulating but also rigorous source that I wish would be essential part of any introductory course into this domain.

Thomas Buhse
Universidad Autónoma del Estado de Morelos
Cuernavaca Morelos, Mexico

FOREWORD II

In his book that covers one academic year of differential and integral calculus for the first-year students, Dr. Carlos Polanco elegantly describes fundamentals of this discipline. The author was able to overcome the usual gap between theory and practice providing in each chapter a multitude of useful examples, exercises, and case studies. This makes the related concepts of Differential Calculus more apprehendable and helps the Reader, even without rigorous mathematical background, to understand these concepts and immediately use them in practice. This book can be used as auxiliary book for students interested in this field, as a reference book for seasoned Researcher, or as a subject refresher for the Researchers in related fields.

Vladimir N. Uversky
University of South Florida
Tampa, Florida, USA

PREFACE

Calculus is an essential mathematical tool for physical and natural phenomena analysis; it is a very important subject for students in the Faculties of Science and Engineering. This book has been designed to cover one academic semester for first-year students, as it contains the fundamentals related to this discipline. Here, the reader will find a concise and clear study of this mathematical field, it provides many examples, exercises, and case studies in each chapter. The complete solutions to the exercises are also included at the end of the book. The reader will be able to understand it, even without rigorous mathematical knowledge and will be able to immediately practice the concepts. Its main purpose is to enable students to put into practice what they are learning about the subject.

This book thoroughly examines the algebraic structure of the **field** and characterizes the set of **real numbers**, as a group formed with the natural, integer, rational, and irrational numbers. It introduces the family of **functions**, their geometric representation, properties, and restrictions. It reviews the concept of **map** in order to show its potential advantages over the functions. The **limit** operator, an essential operator in this discipline, is defined and some examples are given. This operator has a major role in the definition of **continuity** and **differentiability** of a function. With these topics, the characterization of this family of functions is concluded.

With the concept of limits, the **sequences** and **series** are introduced and the techniques to identify their convergence are reviewed. Finally the **composition of functions**, their **implicit differentiation**, as well as Fourier series, and Taylor series are studied.

The author hopes the reader interested in the study of the fundamentals of differential and integral calculus, finds useful the material presented here and that those who start studying this field find this information motivating. The author would like to acknowledge the Faculty of Sciences at Universidad Nacional Autónoma de México for support.

CONSENT FOR PUBLICATION

Not applicable.

CONFLICT OF INTEREST

The author declares no conflict of interest regarding the contents of each of the chapters of this book.

Carlos Polanco
Department of Mathematics
Faculty of Sciences
Universidad Nacional Autónoma de México
México City
México

&

Department of Electromechanical Instrumentation
Instituto Nacional de Cardiología "Ignacio Chávez"
México

ACKNOWLEDGEMENTS

I would like to thank all those whose recommendations made possible the publication of this ebook.

List of Credits

List of Symbols

Symbol	Description	Page
$\mathbb{R}$	Set of Real numbers.	1
$\mathbb{R}^+$	Set of positive real numbers.	2
$\mathbb{N}$	Set of Natural numbers.	3
$\mathbb{Z}$	Set of Integer numbers.	3
$\mathbb{Z}^+$	Set of positive integer numbers.	3
$\mathbb{Q}$	Set of Rational numbers.	4
$\mathbb{I}$	Set of Irrational numbers.	5
(a, b)	Open interval.	5
$[a, b]$	Closed interval.	5
$[a, b)$	Half-open interval.	5
$|a|$	Absolute value of a.	6
$\max A$	Maximum element of set A.	9
$\min A$	Minimum element of set A.	9
$\sup A$	Supremum element of set A.	10
$\inf A$	Infimum element of set A.	10
f	$f: U \subset \mathbb{R} \to M \subset \mathbb{R}$.	14
$f(x)$	Dependent variable of function f.	17
x	Independent variable of function f.	17
D_f	Domain of function.	17
I_f	Image of function.	18
$\text{Graph} f$	Graph of function.	18
$\text{map} T$	Map T.	20
f^{-1}	Inverse function of f.	25
$f \circ g$	Composition of functions.	25
$\sin x$	Sine function.	33
$\cos x$	Cosine function.	34
$\tan x$	Tangent function.	35
$\lim_{x \to x_0} f(x)$	Limit of function.	44
$f'(x)$	Lagrange's notation of derivative f.	75
$(f(x))'$	Notation of derivative.	77
$\dfrac{df}{dx}$	Leibniz's notation of derivative	77
C^1 function	Function of class C^1.	78
$f(x, y)$	Function of two variables.	88
(a_n)	$f: A \subset \mathbb{N} \to S$.	96

CHAPTER 1

Real Number

Abstarct: The purpose of this chapter is to introduce different numeric sets that form the real numbers $\mathbb{R}$, their properties and their geometrical representation in an oriented number line. Some specific properties are addressed, such as irreducible fractions and divisibility.

Keywords: $|a|$, $\mathbb{R}^+$, $\mathbb{Z}^+$, $\mathbb{Z}$, Absolute value, Closed intervals, Half-open intervals, Infimum and Supremum, Integer numbers, Irrational numbers, Maximum element, Minimum element, Natural numbers, Open intervals, Properties of a real field, Rational numbers, Real field $\mathbb{R}$, Triangle inequality.

1.1. LIMITATIONS

This section describes the **field** of real numbers, the subsets that form it and their graphical representation in the **oriented real line**. The large theorems of the properties of the **Real field** are described without proof.

1.2. ALGEBRAIC PROPERTIES OF THE SET $\mathbb{R}$

The **set** named $\mathbb{R}$ is an **infinite** (Def. 1.3), **dense** (Def. 1.5), and **ordered** (Def. 1.4) set that complies with the properties of the **field** [20] and it is formed by all of the numbers in the oriented number line (Fig. **1.1**). The elements of the set $\mathbb{R}$ are called **real** numbers and they are sub-classified into these sets: $\mathbb{N}$ (formed by the natural numbers), $\mathbb{Z}$ (formed by the integer numbers), $\mathbb{Q}$ (formed by the rational numbers), and $\mathbb{I}$ (formed by the irrational numbers), where:

$$\mathbb{N} \subset \mathbb{Z} \subset \mathbb{Q}, \text{and} \mathbb{Q} \cup \mathbb{I} = \mathbb{R}.$$

Definition 1.2. In the set $\mathbb{R}$ is included the subset $\mathbb{R}^+$, defined by $\mathbb{R}^+ = \{x \in \mathbb{R} | \ x \geq 0\}$; that complies with $\mathbb{R}^+ \subset \mathbb{Q} \cup \mathbb{I} = \mathbb{R}$.

The numerical sets forming the $\mathbb{R}$ set, do not meet all of the properties of the **field** (Prop. 1.2). For instance, the set $\mathbb{N}$ does not comply with the 3rd property and set $\mathbb{Z}$ is not dense. However, collectively all these sets meet the properties of the **field**.

Carlos Polanco
All rights reserved-© 2020 Bentham Science Publishers

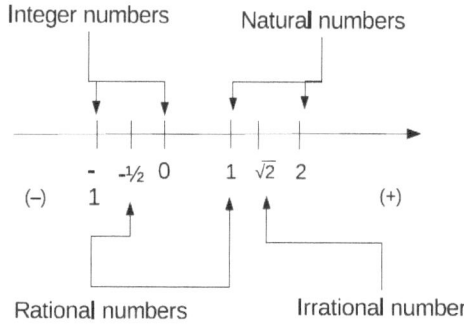

Fig. (1.1). The oriented line is the geometrical representation of the set $\mathbb{R}$.

Definition 1.3. An **infinite** set is a set that has no beginning or end when counting.

Example 1.1. These sets are infinite $\{\cdots,3\}$, $\{0,\cdots\}$, or $\{\cdots,-2,\cdots\}$.

Definition 1.4. In an **ordered** set of two elements a and b, only one of the following affirmations can occur $a < b$, $a > b$, or $a = b$.

Example 1.2. Be the numbers 1 and 2, then $1 < 2$, $1 \neq 2$, or 12.

Definition 1.5. In a **dense** set, for any of the two elements a and b there exists an element x, such that $a < x < b$.

Example 1.3. Be the numbers $\frac{a}{b}$, where $a > b$; the number $\frac{\frac{1}{a}+\frac{1}{b}}{2}$ is always between them.

Property 1. $a + b = b + a, \forall\, a, b \in \mathbb{R}$ (commutative property of addition).
Property 2. $a + (b + c) = (a + b) + c, \forall\, a, b, c \in \mathbb{R}$ (associative property of addition).
Property 3. $\exists!\ 0 \in \mathbb{R}$ such that $0 + a = a$ and $a + 0 = a$, $\forall\, a \in \mathbb{R}$ (existence of a zero element).
Property 4. For each $a \in \mathbb{R}$ there exist an element $-a \in \mathbb{R}$ such that $a + (-a) = 0$ and $(-a) + a = 0$ (existence of negative elements).
Property 5. $a \cdot b = b \cdot a$, $\forall a, b \in \mathbb{R}$ (commutative property of multiplication).
Property 6. $a \cdot (b \cdot c) = (a \cdot b) \cdot c$, $\forall a, b, c \in \mathbb{R}$ (associative property of multiplication).
Property 7. $\exists 1 \in \mathbb{R}$, distinct from 0, such that $1 \cdot a = a$ and $a \cdot 1 = a$, $\forall a \in \mathbb{R}$ (existence of a unit element).

Property 8. $\forall a \in \mathbb{R}, a \neq 0, \; \exists \; \frac{1}{a} \in \mathbb{R}$, such that $a \cdot \left(\frac{1}{a}\right) = 1$ and $\left(\frac{1}{a}\right) \cdot a = 1$
(existence of reciprocals).

Property 9. $a \cdot (b + c) = (a \cdot b) + (a \cdot c)$ and $(b + c) \cdot a = (b \cdot a) + (c \cdot a)$, $\forall a, b, c \in \mathbb{R}$ (distributive property of multiplication over addition).

1.2.1. Natural Numbers

The set named $\mathbb{N}$ is an **infinite** (Def. 1) and **ordered** (Def. 1) set formed by counting cardinal numbers

$$\mathbb{N} = \{1,2,3,\cdots\}.$$

Example 1.4. Are the following sets contained in set $\mathbb{N}$? (i) $\{0,1,3\}$. (ii) $\{1\}$. (iii) $\{-2,3,4\}$. (iv) $\{8,5,11,34\}$.

Solution 1.1. (i) No, it is not. The number $0 \notin \mathbb{N}$. (ii) Yes, it is. The set $\{1\} \subset \mathbb{N}$. (iii) No, it is not. The element $-2 \notin \mathbb{N}$. (iv) Yes, it is. All elements in set $\{8,5,11,34\}$ are natural numbers, so this set is a subset of $\mathbb{N}$.

1.2.2. Integer Numbers

Definition 1.7. The set named $\mathbb{Z}$ is an **infinite** (Def. 1.3) and **ordered** (Def. 1.4) set formed by the union of (i) the set of positive counting or cardinal numbers, (ii) the set of negative counting or cardinal numbers, and (iii) the zero

$$\mathbb{Z} = \{\cdots, -1,0,1,\cdots\}.$$

Definition 1.8. The set $\mathbb{Z}$ contains the subset $\mathbb{Z}^+$ defined by $\mathbb{Z}^+ = \{x \in \mathbb{Z} | \; x \geq 0\}$, as it complies with $\mathbb{N} \cup \{0\} = \mathbb{Z}^+ \subset \mathbb{Z}$.

Theorem 1.1: *(Archimedean Property) If $x > 0$ and $y > 0$ are real numbers, then there exists a positive integer n such that $nx > y$ [21].*

Example 1.5. Are the following sets contained in set $\mathbb{Z}$? (i) $\{0,1,3\}$. (ii) $\{-1\}$. (iii) $\{-2,3,\frac{2}{3}\}$. (iv) $\{8,5,11,\pi\}$.

Solution 1.2. (i) Yes, it is. $\forall x \in \{0,1,3\}, x \in \mathbb{Z}$. (ii) Yes, it is. The set $\{1\} \subset \mathbb{Z}$.
(iii) No, it is not. The element $\frac{2}{3} \notin \mathbb{Z}$. (iv) No, it is not. The element $\frac{2}{3} \notin \mathbb{Z}$, so
$\{8,5,11,\pi\} \not\subset \mathbb{Z}$.

Definition 1.9. For all integer numbers m and $n \in \mathbb{Z}$, m **divides** n if there exists
an integer k such that $mk = n$.

Example 1.6. Does number 3 divide number 6?

Solution 1.3. Yes, it does, because $\exists\, k = 2 \in \mathbb{Z}$ such that $mk = (3)(2) = 6 = n$.

Example 1.7. Does number 2 divide number 3?

Solution 1.4. No, it does not because ó $k \in \mathbb{Z}$ such that $= mk = 2k = 3 = n$.

1.2.3. Rational Numbers

Definition 1.10. The set $\mathbb{Q}$ is an **infinite** (Def. 1.3), **dense** (Def. 1.5), and **ordered**
(Def. 1.4) set, formed by any number that can be expressed as the quotient or
fraction $\frac{p}{q}$ of two integer numbers, a numerator p and a non-zero denominator q
[22]. Every positive rational number can be represented as an **irreducible fraction**
in exactly one way (Def. 1.11). The decimal expansion of a rational number is finite
e.g. 1.3456, or it can be repetitive *e.g.* $1.\overline{3} = 1.3333\cdots$ but its period is **finite**.

Definition 1.11. $\frac{p}{q}$ is irreducible if $\forall p, q \in \mathbb{Z}, q \neq 0 \Leftrightarrow mcd(p,q) = 1$.

Remark 1.1. $\mathbb{N} \subset \mathbb{Z} \subset \mathbb{Q}$.

Theorem 1.2: *(Density of $\mathbb{Q}$): Be $a, b \in \mathbb{R}$ where $a < b$, and $a, b > 0 \ \exists\, m$,
n integers such that $a < \frac{m}{n} < b$ [21].*

Proof. Since $b > a \Rightarrow b - a > 0$ (Thm. 2.2), $\exists\, n \in \mathbb{N}$ such that $n(b - a) > 1 \Rightarrow$
$bn - an > 1$. Thus, there exist $m, n \in \mathbb{N}$ such that $an < m < bn$, which implies
$a < m/n < b$.

Example 1.8. Are the following sets contained in $\mathbb{Q}$? (i) $\{0,1,3\}$. (ii) $\{-\frac{1}{2}\}$. (iii) $\{-2,3,\sqrt{2}\}$.

Solution 1.5. (i) Yes, it is. $\forall x \in \{0,1,3\}, x \in \mathbb{Z}$, and $\mathbb{Z} \subset \mathbb{Q}$. (ii) Yes, it is. The element $\frac{1}{2} \in \mathbb{Q}$. (iii) No, it is not. The element $\sqrt{2} \notin \mathbb{Q}$ (Prf. 1.5).

Proof. **Suppose** $\sqrt{2}$ is **a rational number**, it can be denoted $\sqrt{2} = \frac{p}{q}$, where $p, q \in \mathbb{Z}, q \neq 0$ and $\frac{p}{q}$ is an irreducible fraction. Then $2 = \frac{p^2}{q^2} \Rightarrow p^2 = 2q^2$, so p^2 is even and p is even. In other words, p has the form $2k$. If we substitute $p = 2k$ into the original equation $2 = \frac{p^2}{q^2} \Rightarrow q^2 = 2k^2$, we can see that q is also even, but $\frac{p}{q}$ was defined as an irreducible fraction; therefore, there is **not** an irreducible fraction for $\sqrt{2}$. So $\sqrt{2}$ is not a rational number.

1.2.4. Irrational Numbers

Definition 1.12. The set $\mathbb{I}$ is an **infinite** (Def. 1.3), **dense** (Def. 1.5), and **ordered** (Def. 1.4) set. It is a number that **cannot** be expressed as a quotient or fraction $\frac{p}{q}$ of two integer numbers, a numerator p and a non-zero denominator q [22]. Therefore, it **cannot** be represented as an **irreducible fraction** in exactly one way (Def. 2.3). The decimal expansion of an irrational number is infinite *e.g.* $\sqrt{7} = 2.645751311064591\cdots$, so its period is **infinite**.

Remark 1.2. Since $\sqrt{2} - \sqrt{2} = 0 \in \mathbb{Q}$, then the sum of two irrationals may be rational [23].

Theorem 1.3. *(Density of $\mathbb{I}$): Be $a, b \in \mathbb{R}$ where $a < b$ and $a, b > 0$. $\exists \ q \in \mathbb{Q} - \mathbb{I}$, such that $a < q < b$ [24].*

Proof. Consider the real numbers $\frac{a}{\sqrt{2}}$ and $\frac{b}{\sqrt{2}}$, then $\exists \ p$ such that $\frac{a}{\sqrt{2}} < p \frac{b}{\sqrt{2}}$ (Thm. 2.3) $\Rightarrow a < \sqrt{2}p < b$. If we take $q = \sqrt{2}p$, q is an irrational number.

Example 1.9. Rationalize $\frac{2}{\sqrt{11}-\sqrt{6}}$.

Solution 1.6. $\frac{2}{\sqrt{11}-\sqrt{6}} \times \frac{\sqrt{11}+\sqrt{6}}{\sqrt{11}+\sqrt{6}} = \frac{2(\sqrt{11}+\sqrt{6})}{11-6} = \frac{2(\sqrt{11}+\sqrt{6})}{5}$.

Example 1.10. Denote in the form $\frac{p}{q}$ the number $0.7\overline{234}$ [25].

Solution 1.7. $x = 0.7234, 10^4 x = 7234.234, \Rightarrow 10^4 x - 10x = 7234.234 - 7.234, \Rightarrow 9990x = 7227, \Rightarrow x = \frac{7227}{9990} = 0.723423423 = 0.7\overline{234}$.

1.2.5. Open, Closed, and Half-Open Intervals

It is possible to define **intervals** in the real line, they can be open intervals, closed intervals, or half-open intervals.

Definition 1.13. An **open interval** $(a, b) \in \mathbb{R}$ is defined by $(a, b) = \{x \in \mathbb{R} | \, a < x < b\}$ (Fig. **1.2**).

Definition 1.14. A **closed interval** $[a, b] \in \mathbb{R}$ is defined by $[a, b] = \{x \in \mathbb{R} | \, a \leq x \leq b\}$ (Fig. **1.2**).

Definition 1.15. A **half-open interval** $[a, b) \in \mathbb{R}$ –or $(a, b]$ is defined by $[a, b) = \{x \in \mathbb{R} | \, a \leq x < b\}$ (Fig. **1.3**), or alternatively $(a, b] = \{x \in \mathbb{R} | \, a < x \leq b\}$ (Fig. **1.3**)

Example 1.11. Is the interval $|x - 1| < 2$ a closed interval?

Solution 1.8. $|x - 1| < 2 \Leftrightarrow -2 < x - 1 < 2 \Leftrightarrow -1 < x < 3$. No, it is not. It is an open interval.

Remark 1.3. It is possible to define the set $\mathbb{R}$ as an open interval $(-\infty, \infty)$. The previous interval definition accepts this type of combinations.

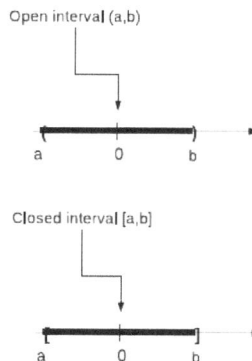

Fig. (1.2). Geometrical interpretation of an open interval (a, b) and a closed interval $[a, b]$.

Half-open interval [a,b)

a   0   b

Half-open interval (a,b]

a   0   b

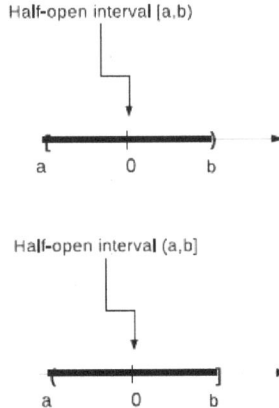

Fig. (1.3). Geometrical interpretation of half-open intervals $[a, b)$ or $(a, b]$.

1.2.6. Absolute Value

Definition 1.16. For any real number x the **absolute value** is denoted by (Eq. 1.1).

$$|x| = \begin{cases} x & : & x \geq 0 \\ -x & : & \text{otherwise} \end{cases} \tag{1.1}$$

1.2.7. Algebraic Properties of the Set $\mathbb{R}$

The **absolute value** operator of a number represents the distance from the origin 0 (Fig. **1.4**), in the case $|a - b|$, it represents the distance between a and b (Fig. **1.5**). Equivalently $|x| = \sqrt{x^2}$. Its main properties are (Prop. 1.2.6).

Property 1. $|x| \geq 0, \forall \, x \in \mathbb{R}$.
Property 2. $|x| = 0 \Leftrightarrow x = 0, \forall \, x \in \mathbb{R}$.
Property 3. $|xy| = |x||y|, \forall \, x, y \in \mathbb{R}$.
Property 4. $|\frac{x}{y}| = \frac{|x|}{|y|}, \forall \, x, y \in \mathbb{R}$, if $y \neq 0$.
Property 5. $|x| \leq a \Leftrightarrow -a \leq x \leq a, \forall \, x, \in \mathbb{R}$ and $a = cte$.
Property 6. $|x| \geq a \Leftrightarrow x \leq -a \, or \, x \leq a, \forall \, x, \in \mathbb{R}$ and $a = cte$.
Property 7. $|x + y| \leq |x| + |y|, \forall \, x, y \in \mathbb{R}$ (triangle inequality).

Example 1.12. Show the triangle inequality [26].

Proof. If (i) $-|x| \leq x \leq |x|$ and (ii) $-|y| \leq y \leq |y|$, then adding (i) and (ii) $-(|x| + |y|) \leq x + y \leq |x| + |y| \Rightarrow |x + y| \leq |x| + |y|$.

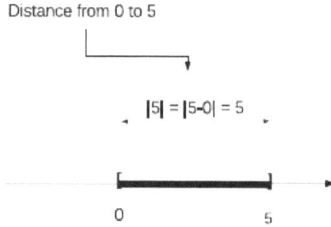

Distance from 0 to 5

$|5| = |5\text{-}0| = 5$

0   5

Fig. (1.4). Geometrical interpretation of $|5|$.

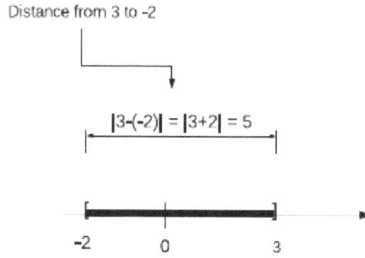

Distance from 3 to -2

$|3\text{-}(\text{-}2)| = |3\text{+}2| = 5$

-2   0   3

Fig. (1.5). Geometrical interpretation of $|3 - (-2)|$.

Example 1.13. Find the solution of $|2x + 3| > 4$, where $x \in \mathbb{R}$.

Solution 1.9. (i) $2x + 3 > 4 \Leftrightarrow 2x > 1 \Leftrightarrow x > \frac{1}{2}$. (ii) $2x + 3 < -4 \Leftrightarrow 2x < -7 \Leftrightarrow x < -\frac{7}{2}$ (Fig. **1.6**).

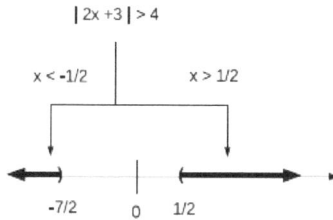

$| 2x +3 | > 4$

$x < \text{-}1/2$ $x > 1/2$

-7/2   0   1/2

Fig. (1.6). Geometrical solution of $|2x + 3| > 4$.

Example 1.14. Show that $|-x| = |x|, \forall\, x \in \mathbb{R}$.

Solution 1.10. $|-x| = \sqrt{(-x)^2} = \sqrt{x^2} = |x|$.

1.2.8. Infimum and Supremum

Definition 1.17. Be a non-empty set $A \subset \mathbb{R}$, a number M is the **maximum** (or **max**) of set A if:

Property 1. $M \in a$,
Property 2. $\forall x \in A, M \geq x$.

If M complies with (Prop. 1.17), then $M = $ **max** A.

Definition 1.18. Be a non-empty set $A \subset \mathbb{R}$, a number m is the **minimum** (or **min**) of set A if:

Property 1. $m \in a$,
Property 2. $\forall x \in A, m \leq x$.

If m complies with (Prop. 2.7), then $m = $ **min** A.

Example 1.15. Be the set $A = \{3,2,1\}$. Determine the **max** A and **min** A elements.

Solution 1.11. (i) $3 \in A \subset \mathbb{R}$. (ii) $3 \geq 1$ and $3 \geq 2$, then $3 = $ **max** A. Similarly the **min** is 1.

Example 1.16. Be the set $\mathbb{N}$. Determine the **max** $\mathbb{N}$ and **min** $\mathbb{N}$.

Solution 1.12. There is no **max** element in $\mathbb{N}$ as it is infinite. The *min* element is 0.

Example 1.17. Be the set $A = (0,1)$. Determine the **max** A [25].

Solution 1.13. This is an open interval, therefore, any element can be surpassed by other in both directions. Consequently, there is neither **max** nor **min**.

Example 1.18. Be the set $A = \{x \in \mathbb{R} \mid x^2 \leq 4\}$. Determine the **max** A and **min** A.

Solution 1.14. The **max** A is 2. The **min** A is -2.

Definition 1.19. Be the closed interval $[a, b] \subset \mathbb{R}$, element x is the **upper bound** of $[a, b]$ if $x \leq b$. Similarly, element x is the **lower bound** of $[a, b]$ if $x \geq a$ (Fig. **1.7**).

Remark 1.4. The **upper bound** or **lower bound** elements do not necessarily have to be part of the interval.

All **upper bounds** of $[a, b]$ are included in the interval $(b, +\infty)$ and all **lower bounds** of $[a, b]$ are included in the interval $(-\infty, a)$ (Fig. **1.7**). The min $(b, +\infty)$ is the **supremum** (or **sup**) element of $[a, b]$ and the max $(-\infty, a)$ is the **infimum** (or **inf**) element of $[a, b]$.

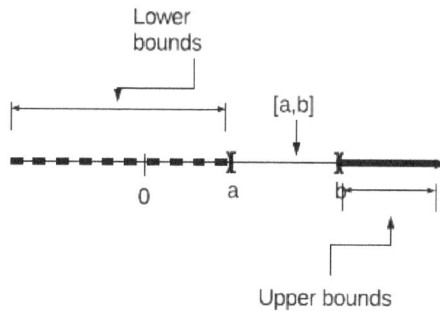

Fig. (1.7). Set of **upper bounds** and **lower bounds** associated with a closed interval $[a, b]$.

Example 1.19. Determine the **lower bounds**, **upper bounds**, **infimum**, and **supremum** elements of the interval $(-2 - 1)$.

Solution 1.15. The **lower bounds** is the interval $(-\infty, -2]$, the **upper bounds** is the interval $[-1, \infty)$, the **infimum** element is -2, and the **supremum** element is -1.

1.3. CASE STUDY: ABSOLUTE VALUE INEQUALITIES

Case 1.1. Determine the value of x that satisfies $|4 + \frac{1}{x}| \leq 4$. The solution of this inequality implies the solution of two cases (Eqs.1.2 and 1.3).

$$-4 \leq \quad 4 + \frac{1}{x} \tag{1}$$

$$4 + \frac{1}{x} \leq \quad 4 \tag{2}$$

Solving (Eq. 1.2) let $x > 0$.

$$-4 \leq \qquad 4 + \frac{1}{x}$$

$$-4 - 4 \leq \quad \frac{1}{x} \quad (\text{if } x > 0) \tag{3}$$

$$-8x \leq \qquad 1$$

$$x \geq \qquad -\frac{1}{8}.$$

From (Eq. 1.4) we have $x > 0$ and $x \geq -\frac{1}{8}$, then $x > 0$.

Now, solving (Eq. 1.2) let $x < 0$ (see Eq. 1.5).

$$-4 \leq \qquad 4 + \frac{1}{x}$$

$$-4 - 4 \leq \quad \frac{1}{x} \quad (\text{if } x < 0) \tag{4}$$

$$-8x \geq \qquad 1$$

$$x \leq \qquad -\frac{1}{8}.$$

From (Eq. 1.5) we have $x < 0$ and $x \leq -\frac{1}{8}$, then $x \leq -\frac{1}{8}$.

Solving the second inequality (Eq. 1.3) in a similar way $x > 0$.

$$4 + \frac{1}{x} \leq \quad 4$$

$$\frac{1}{x} \leq \qquad 4 - 4 \quad (\text{if } x > 0) \tag{5}$$

(Eq. 1.6) is false for $x > 0$, now we solve for (Eq. 1.3) taking $x < 0$ (Eq. 1.7), to see if it is true.

$$4 + \frac{1}{x} \leq 4$$

$$\frac{1}{x} \geq 4 - 4 \quad (\text{if } x < 0) \tag{6}$$

(Eq. 1.7) is always true.

Putting together the solutions for each case we have $x \in (-\infty, -\frac{1}{8}]$.

1.4. CASE STUDY: RATIONALIZATION OF IRRATIONAL NUMBERS

Case 1.2. Rationalization is a mathematical procedure to suppress radicals in the denominator of a rational number p/q. The procedure consists of multiplying p/q by a unit r, this eliminates the denominator of the radical (Eq. 1.9). Let's see the following case.

$$\frac{1}{\sqrt[5]{x^3}} = \left[\frac{1}{\sqrt[5]{x^3}}\right]\left[\frac{\sqrt[5]{x^2}}{\sqrt[5]{x^2}}\right]$$

$$= \frac{\sqrt[5]{x^2}}{\sqrt[5]{(x^2)(x^3)}}$$

$$= \frac{\sqrt[5]{x^2}}{\sqrt[5]{x^5}} \tag{7}$$

$$= \frac{\sqrt[5]{x^2}}{x}.$$

Here, you can see how the numerator and the denominator are multiplied by the same number, in such a way that it equals the exponent of the root in the denominator, canceling it and taking it to the numerator without changing the result.

Getting the value of a from (Eq. 1.9).

$$\frac{1}{\sqrt{17}-\sqrt{15}} = \frac{\sqrt{a}+\sqrt{15}}{2} \tag{8}$$

First, we simplify the term at the left of (Eq. 1.9),

$$\frac{1}{\sqrt{17}-\sqrt{15}} = \frac{\sqrt{17}+\sqrt{15}}{2} \tag{9}$$

Then, comparing the term at the right (Eq. 1.9) with the term at the left (Eq. 1.10), we obtain $a = 17$.

$$\frac{\sqrt{17}+\sqrt{15}}{2} = \frac{\sqrt{a}+\sqrt{15}}{2} \tag{10}$$

1.5. CASE STUDY: CONSTRUCTION OF IRRATIONAL NUMBERS

Case1.3. To construct an irrational number $\sqrt{c}$ on the oriented real line, we have to construct a triangle whose sides a and b comply with $c^2 = a^2 + b^2$ on the oriented real line (Fig. **1.8**). Observe that the diagonal line of the rectangle represents the irrational number $\sqrt{c}$. Now, translate this diagonal line with a compass to the oriented real line.

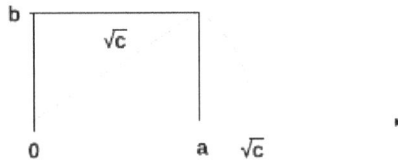

Fig. (1.8). Construction of an irrational number $\sqrt{c}$ using a rectangle on the oriented real line.

1.6. EXERCISES

Exercise 1.1. Draw the following intervals $(-\infty, -3)$, $[-2,0]$, and $[1, \infty)$, on the real oriented line.

Exercise 1.2. Demonstrate that $a + b = a + c \Rightarrow b = c$? for all real numbers. **Hint**, use (Prop. 1.2).

Exercise 1.3. Does number -2 divide number 4? Explain the concept of divisibility and provide an example showing the contrary.

Exercise 1.4. Find the set of real numbers that solves the equation $|\frac{x-3}{x+2}| = 2$. **Hint**, use $|x| = a \Leftarrow -a = x = a$.

Exercise 1.5. Find the set of real numbers that solves the equation $\sqrt{2x + 2} + x = 1$.

Exercise 1.6. Find the set of real numbers that solves the inequality $10x + 5 \leq 25$.

Exercise 1.7. Find the set of real numbers that solves the inequality $7 \leq 3x - 2 \leq 13$.

Exercise 1.8. Find the set of real numbers that solves the inequality $\frac{3}{x} < 5, x \neq 0$.

Exercise 1.9. Find the set of real numbers that solves the inequality $\frac{2x+1}{3x-6} \geq 3$. **Hint,** use (Prop. 2.6–6).

Exercise 1.10. Find the set of real numbers that solves the inequality $x^2 + x > 6$. **Hint,** Use $ab > 0$ if both terms are positive or negative.

Exercise 1.11. Determine the **lower bounds, upper bounds, infimum** and **supremum** elements of the set $A = \{-6, -3, 0, 3, 6\}$.

Exercise 1.12. Determine the **lower bounds, upper bounds,** the **infimum** and the **supremum** elements of the set $A = \{\cdots, -5, -2, 0, 1, 4, 8\}$.

Exercise 1.13. Determine the **lower bounds, upper bounds, infimum** and **supremum** elements of the set $A = (-3, 4]$.

Exercise 1.14. Show that if $|x - 2| < 1$, then $|x^2 - 4| < 5$?. **Hint,** compare the interval that is the solution with both inequalities.

Exercise 1.15. Determine the elements in the set $\mathbb{R}$ that satisfy $|x - 1| < |x|$. **Hint,** use $|x| = \sqrt{x^2}$ in both terms.

Exercise 1.16. Show the following equality $|x| = |-x|, \forall x \in \mathbb{R}$. **Hint,** Review cases $x \geq 0$ and $x < 0$.

Exercise 1.17. Find the set of real numbers that solves the inequality $|5x - 1| < \frac{3}{5}$.

CHAPTER 2

Functions and Maps

Abstarct: This section will review the functions and maps operators only using the regularities and differences found in their graphics. A broader description will be presented in (Chap. 5), once the concepts of limit (Chap. 3), continuity (Chap. 4), and differentiability (Chap. 5) are studied.

Keywords: Algebraic functions and maps, Bijective functions and maps, Classification of functions and maps, Composition of functions and maps, Cosine function and cosine map, Domain of a map, Domain of a function, Elementary functions, Elementary maps, Explicit functions, Exponential function, Exponential map, Functions, Graph of a map, Graph of a function, Image of a map, Image of a function, Injective functions, Injective maps, Inverse functions, Inverse maps, Irrational functions, Irrational maps, Linear functions, Linear maps, Logarithmic function, Logarithmic map, Maps, Maps $\mathbb{R} \to \mathbb{R}$, Maps $\mathbb{R} \to \mathbb{R}^2$, Non-numeric functions, Polynomial equation, Quadratic functions, Quadratic maps, Rational functions, Rational maps, Sine function, Sine maps, Surjective functions, Surjective maps, Tangent function, Tangent maps, Transcendental functions, Transcendental maps.

2.1. FUNCTIONS

A **function**, in a general sense, is a **well defined rule** f that relates two sets $f: A \to B$, where each element in set A corresponds to a **unique** element in set B.

Example 2.1. Let the **non-numeric function** $f: \mathbb{A} \to \mathbb{B}$ allocates the gender "male" to "boys" or "female" to "girls". (i) Obtain $f(x)$ for the elements $x = \{\text{boy}, \text{girl}\}$. (ii) Is f a function?

Solution 2.1. (i) $f(\text{boy}) = \text{male}$, $f(\text{girl}) = \text{female}$. (ii) Yes it is, f is a function because its rule is well defined.

Example 2.2. Let the non-numeric function $f: \mathbb{A} \to \mathbb{B}$ allocates the gender "male" or "female", without specifying who the gender is allocated to. (i) Obtain $f(x)$ for the elements $x = \{\text{boy}, \text{girl}\}$. (ii) Is f a function?

Solution 2.2. (i) $f(\text{boy}) = \text{male or female}$, $f(\text{girl}) = \text{maleor female}$. (ii) Not it is not, f is **not** a function because its rule is **not** well defined.

Carlos Polanco
All rights reserved-© 2020 Bentham Science Publishers

Definition 2.1. A function f is a **rule** $f: U \subset \mathbb{R} \to M \subset \mathbb{R}$, [rule] such that for each element $x \in U$ a **unique** element $f(x) \in M$ [20] is assigned.

Remark 2.1. The element x is the **independent** variable [28] and the element $f(x)$ is the **dependent** variable [28].

Remark 2.2. In this book we will preferably use the notation $f: U \subset \mathbb{R} \to M \subset \mathbb{R}$[rule] for the function and in the cases where there is no possibility for confusion we will use $f(x) = $ [rule].

Exmaple 2.3. Is the equation $f(x) = \pm\sqrt{1 - x^2}$ a function?

Solution 2.3. No, it is not because for any element $x \in (-1,1)$ corresponds two values, one positive and one negative.

The real valued functions meet (Prop. 6.2)

Property 1. $(f + g)(x) = f(x) + g(x)$ (addition).
Property 2. $(f - g)(x) = f(x) - g(x)$ (substraction).
Property 3. $(fg)(x) = f(x) \cdot g(x)$ (multiplication).

Example 2.4. Let the function $f: \mathbb{R} \Rightarrow \mathbb{R}, x^2 + 3$. Obtain $f(x)$ for the elements of $x = \{-3, -1, 0\}$.

Solution 2.4. $f(-3) = (-3)^2 + 3 = 12$, $f(-1) = (-1)^2 + 3 = 4$, and $f(0) = (0)^2 + 3 = 3$.

Example 2.5. Let the function $f: \mathbb{R} \Rightarrow \mathbb{R}, x^2 + 3$ and $f: \mathbb{R} \Rightarrow \mathbb{R}, x + 1$. Obtain (i) $(f + g)(x)$. (ii) $(f - g)(x)$. (iii) $(fg)(x)$.

Solution 2.5. (i) $(f + g)(x) = f(x) + g(x) = x^2 + 3 + x + 1 = x^2 + x + 4$. (ii) $(f - g)(x) = f(x) - g(x) = x^2 + 3 - x - 1 = x^2 - x + 2$. (iii) $(fg)(x) = f(x) \cdot g(x) = (x^2 + 3)(x + 1) = x^3 + x^2 + 3x + 3$.

2.1.1. Domain of a Function

Definition 2.2. The domain D_f (Eq. 2.1) of a function $f: U \subset \mathbb{R} \to M \subset \mathbb{R}$ is the set U, so the domain D_f is defined

$$D_f = \{x \in U \subset \mathbb{R} \mid \exists\, f(x) \in M \subset \mathbb{R}\} \tag{2.1}$$

Example 2.6. Be $f(x) = x^2$. (i) Describe this function using an equivalent notation. (ii) What is the domain of f?

Solution 2.6. (i) $f: U \in \mathbb{R} \to \mathbb{R}^+, x^2$ (Sect. 2.2). (ii) The domain is $\mathbb{R}$.

2.1.2. Image of a Function

Definition 2.3. The image I_f (Eq. 2.2) of a function $f: U \subset \mathbb{R} \to M \subset \mathbb{R}$ is the set $M \subset R$.

$$I_f = \{y \in M \subset \mathbb{R} \mid y = f(x), \text{for some } x \in U \subset \mathbb{R}\} \tag{2.2}$$

Example 2.7. Be the function $f: U \subset \mathbb{R} \to M \subset \mathbb{R}, x$. (i) What is the domain? (ii) What is the image?

Solution 2.7. (i) The domain D_f is the space $\mathbb{R}$. (ii) The image I_f is the space $\mathbb{R}$.

Example 2.8. Be $f(x) = \frac{4x^2+6}{x}$. (i) Where is located the domain of f? (ii) Where is located the image of f?

Solution 2.8. (i) The domain is located in $\mathbb{R} - \{0\}$. (ii) The image is located in $[6, \infty) \subset \mathbb{R}^+$.

Example 2.9. Be $f(x) = \sqrt{x-4}$. (i) Describe this function using an equivalent notation. (ii) What is the domain of f? (iii) What is the image of f?

Solution 2.9. (i) $f: [4, \infty] \in \mathbb{R}^+ \to \mathbb{R}^+, \sqrt{x-4}$. (ii) The domain is $[4, \infty] \subset \mathbb{R}^+$. (iii) The image is $\mathbb{R}^+$.

2.1.3. Graph of a Function

Definition 2.4. The graph of a function $f: U \subset \mathbb{R} \to M \subset \mathbb{R}$ (Eq. 2.16) is the geometric representation of the ordered pair $(x, f(x)) \in \mathbb{R}^2$.

$$\text{Graph } (f) = \{(x, f(x)) \in \mathbb{R}^2 \mid x \in U\} \tag{2.3}$$

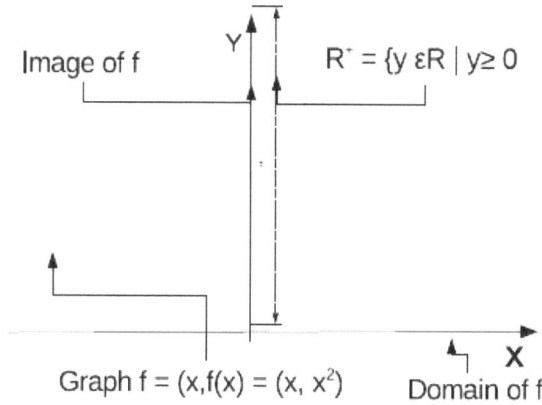

Fig. (2.1). The graph of function $f: \mathbb{R} \to \mathbb{R}, x^2$ is the set of points $(x, f(x)) = (x, x^2) \in \mathbb{R}^2$, its domain D_f is the x-axis $\mathbb{R}$, and its image I_f is the $\mathbb{R}^+$ in the y-axis.

Example 2.10. Be the function $f(x) = x^2$. Draw the graph of f and points D_f and I_f.

Solution 2.10. The graph of function f is a parabola oriented upward from the origin around the y-axis (Fig. **2.1**), its domain D_f is the set $\mathbb{R}$ located on the x-axis, and its image I_f is the set $\mathbb{R}^+$ located on the y-axis.

2.2. MAPS

A map T acts as a function f regardless of the rule that for each element $x \in U$ there is a **unique** element, such that $f(x) \in M$.

Definition 2.5. A **map** T is a **rule** $f: U \subset \mathbb{R} \to M \subset \mathbb{R}$, [rule] such that for each element $t \in U$ an element $T(t) \in M$ is assigned.

Remark 2.3. The definition of a function (Def. 2.1) is more restrictive than the definition of a map (Def. 2.5), because a map does not require a **unique** element in its domain is assigned to an element of its image.

In the first chapters Chap. 2-4 we will study only the maps T described below, and in Chap. 5 we will introduce a variation of map T.

1. Maps $T: U \subset \mathbb{R} \to M \subset \mathbb{R}, (T_1)$, transform the points $t \in U$, located in a real line $\mathbb{R}$, in points of the form $T_1 \in M$, also located in a real line. Where T_1 is the short notation of a real function $T_1: \mathbb{R} \to \mathbb{R}$.
2. Maps $T: U \subset \mathbb{R} \to M \subset \mathbb{R}^2, (T_1, T_2)$, transform the points $t \in U$, in points of the form $(T_1, T_2) \in M$, that are ordered pairs located in a Cartesian plane $\mathbb{R}^2$. Where T_1, and T_2 are the short notation of the real functions $T_1: \mathbb{R} \to \mathbb{R}$, and $T_2: \mathbb{R} \to \mathbb{R}$.

Example 2.11. Classify, according to the two types of maps, the following expressions: (i) $T(t) = (t)$. (ii) $T(t) = \sin t$. (iii) $T(t) = (t, \cos t)$, (iv) $T(t) = (2t, 3\sin t)$.

Solution 2.11. (i) It is a map type 1. (ii) This expression is not denoted correctly. (iii) It is a map type 2. (iv) It is a map type 2.

2.2.1. Domain of a Map

Definition 2.6. The domain D_T (Eq. 2.4) of a map $T: U \subset \mathbb{R} \to M \subset \mathbb{R}, (T_1)$ is the set U, so the domain D_T is defined

$$D_T = \{t \in U \subset \mathbb{R} \mid \exists\ T_1(t) \in M \subset \mathbb{R}\} \tag{2.4}$$

Definition 2.7. The domain D_T (Eq. 2.5) of a map $T: U \subset \mathbb{R} \to M \subset \mathbb{R}^2, (T_1, T_2)$ is the set U, so the domain D_T is defined

$$D_T = \{t \in U \subset \mathbb{R} \mid \exists\ (T_1(t), T_2(t)) \in M \subset \mathbb{R}^2\} \tag{2.5}$$

Example 2.12. Be $T(t) = (t^2, 2t)$. (i) Describe this map using an equivalent notation. (ii) What is the domain of T?

Solution 2.12. (i) $T: U \in \mathbb{R} \to \mathbb{R}^2 (t^2, 2t)$. (ii) The domain is $\mathbb{R}$.

2.2.2. Image of a Map

Definition 2.8. The image I_T (Eq. 2.6) of a map $T: U \subset \mathbb{R} \to M \subset \mathbb{R}, (T_1)$ is the set $M \subset R$.

$$I_T = \{T_1(t) \in M \subset \mathbb{R} \mid \text{for some } t \in U \subset \mathbb{R}\} \tag{2.6}$$

Definition 2.9. The image I_T (Eq. 2.7) of a map $T: U \subset \mathbb{R} \to M \subset \mathbb{R}^2, (T_1, T_2)$ is the set $M \subset R$.

$$I_T = \{(T_1(t), T_2(t)) \in M \subset \mathbb{R} \mid \text{for some } t \in U \subset \mathbb{R}\} \tag{2.7}$$

Example 2.13. Be the map $T: U \subset \mathbb{R} \to M \subset \mathbb{R}, (t)$. (i) What is the domain? (ii) What is the image?

Solution 2.13. (i) The domain D_T is the space $\mathbb{R}$. (ii) The image I_T is the space $\mathbb{R}$.

2.2.3. Graph of a Map

Definition 2.10. The graph of a map $T: U \subset \mathbb{R} \to M \subset \mathbb{R}, (T_1(t))$ (Eq. 2.9) is the geometric representation of the points $T_1(t) \in M \subset \mathbb{R}$.

$$\text{Graph } (T) = \{T_1(t) \in \mathbb{R} \mid t \in U\} \tag{2.8}$$

Definition 2.11. The graph of a map $T: U \subset \mathbb{R} \to M \subset \mathbb{R}, (T_1, T_2)$ (Eq. 2.9) is the geometric representation of the points $(T_1, T_2) \in M \subset \mathbb{R}^2$.

$$\text{Graph } (T) = \{(T_1(t), T_2(t)) \in \mathbb{R}^2 \mid t \in U\} \tag{2.9}$$

2.2.4. Maps $\mathbb{R} \to \mathbb{R}$

Definition 2.12. A map T has the rule $T: [t_1, t_2] \subset \mathbb{R} \to \mathbb{R}, (T_1(t))$. Thus, $\forall\ t \in D_T$ implies $T_1(t) \in I_T$. (Fig. **2.2**) [29].

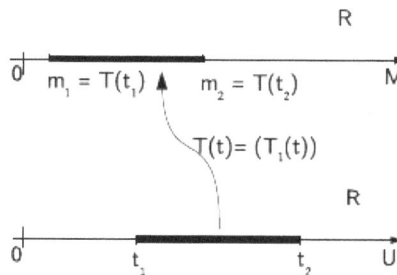

Fig. (2.2). Map $T: [t_1, t_2] \subset \mathbb{R} \to [t_1, t_2] \subset \mathbb{R}$, where $t = T_1(t)$.

Remark 2.4. This map transforms lines into lines.

Example 2.14. Be the map $T: [0,3] \subset \mathbb{R} \to [0,9] \subset \mathbb{R}, (t^2)$. (i) What is the domain of T? (ii) What is the image of T?

Solution 2.14. (i) The domain of T is the closed interval $[0,3]$. (ii) The image of T is the closed interval $[0,9]$.

2.2.5. Maps $\mathbb{R} \to \mathbb{R}^2$

Definition 2.13. A map T has the rule $T: t \subset \mathbb{R} \to \mathbb{R}^2, (T_1(t), T_2(t))$. Thus, $\forall\ t \in D_T$ implies $(T_1(t), T_2(t)) \in I_T$. (Fig. **2.3**) [29].

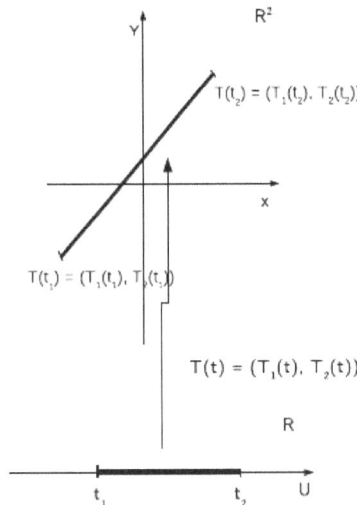

Fig. (2.3). Map $T: t \in \mathbb{R} \to \subset \mathbb{R}^2, (T_1(t), T_2(t))$.

Remark 2.5. This map transforms lines into curves in the plane.

Example 2.15. Be the map $T: [0,2\pi] \subset \mathbb{R} \to \mathbb{R}^2, (\cos t, \sin t)$. (i) What is the domain of T? (iii) What is its graph?

Solution 2.15. (i) The domain of T is the closed interval $[0,2\pi]$. (ii) Its graph is a circumference, not a solid circle. See (Fig. **2.4**)

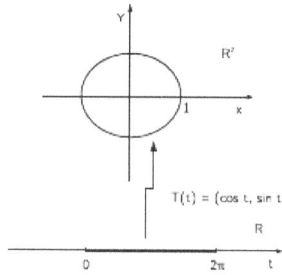

Fig. (2.4). Map $T: [0, 2\pi] \subset \mathbb{R} \to [-1, 1] \subset \mathbb{R}^2$, $(\cos t, \sin t)$. Figure taken from [29].

2.3. CLASSIFICATION OF FUNCTIONS AND MAPS

A function f or map T can be injective, surjective, or bijective depending on the association of the elements of its **domain** and its **image**.

2.3.1. Injection Type

Definition 2.14. A function $f: U \subset \mathbb{R} \to M \subset \mathbb{R}$ is **injective** (Fig. **2.5**) if f meets (Eq. 2.10).

$$\forall \ x_1, x_2 \in U, f(x_1) = f(x_2) \Rightarrow x_1 = x_2 \Leftarrow f(x_1) \neq f(x_2) \Rightarrow x_1 \neq x_2. \quad \textbf{(2.10)}$$

Definition 2.15. A map $T: U \subset \mathbb{R} \to M \subset \mathbb{R}$, $(T_1(t))$ is **injective** if T meets (Eq. 2.11).

$$\forall \ t_1, t_2 \in U, T_1(t_1) = T_2(t_2) \Rightarrow t_1 = t_2 \Leftarrow T_1(t_1) \neq T_1(t_2) \Rightarrow t_1 \neq t_2. \quad \textbf{(2.11)}$$

Example 2.16. (i) Is function $f: \mathbb{R} \to \mathbb{R}, x^2$ an injective function? (ii) Is function $f: [0, \infty) \subset \mathbb{R} \to \mathbb{R}^+, x^2$ an injective function?

Solution 2.16. (i) No, it is not because $f(-1) = f(1)$ and $-1 \neq 1$. (ii) Yes, it is because $\forall \ x_1, x_2 \in [0, \infty), f(x_1) \neq f(x_2) \Rightarrow x_1 \neq x_2$.

Example 2.17. Proof that $f(x) = 4x + 1$ is an injective function.

Solution 2.17. Let $f: U \subset \mathbb{R} \to M \subset \mathbb{R}$, suppose $f(x_1) = f(x_2)$, so $4x_1 + 1 = 4x_2 + 1 \to 4x_1 = 4x_2 \to x_1 = x_2$. According to definition (Eq. 2.10) f is injective.

$$f(x_1) \neq f(x_2) \rightarrow x_1 \neq x_2$$

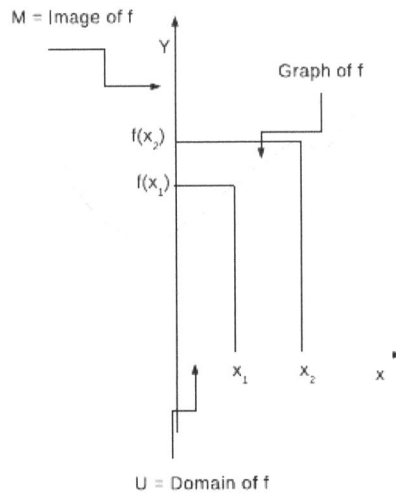

M = Image of f

Y

Graph of f

$f(x_2)$

$f(x_1)$

x_1 x_2 x

U = Domain of f

Fig. (2.5). An injective function must satisfy $f(x_1) \neq f(x_2) \Rightarrow x_1 \neq x_2$.

2.3.2. Surjection Type

Definition 2.16. A function $f: U \subset \mathbb{R} \rightarrow M \subset \mathbb{R}$ is **surjective** (Fig. **2.6**) if f meets (Eq. 2.12).

$$\forall \; y \in M, \; \exists \, x \in U \text{ such that } f(x) = y. \tag{2.12}$$

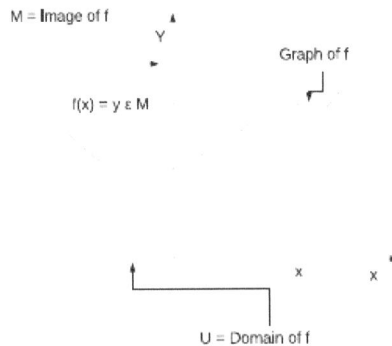

M = Image of f

Y

Graph of f

$f(x) = y \, \varepsilon \, M$

x   x

U = Domain of f

Fig. (2.6). A surjective function must satisfy that there exist $x \in U$ such that $f(x) = y$.

Definition 2.17. A map $T: U \subset \mathbb{R} \to M \subset \mathbb{R}, T_1(t)$ is **surjective** if T meets (Eq. 2.13).

$$\forall \ y \in M, \ \exists \ t \in U \text{ such that } T_1(t) = y. \tag{2.13}$$

Example 2.18. (i) Is function $f: U \subset \mathbb{R} \to M \subset \mathbb{R}, x^2$ a surjective function? (ii) Is function $f: U \subset \mathbb{R}^+ \to M \subset \mathbb{R}^+, x^2$ a surjective function?

Solution 2.18. (i) No, it is not because $-1 \in M \subset \mathbb{R}$, but ó $x \in M \subset \mathbb{R}$ such that $f(x) = -1$. (ii) Yes, it is because $\forall \ y \in M \subset \mathbb{R}^+, \ \exists \ x \in U \subset \mathbb{R}^+$.

Example 2.19. Proof that $f(x) = 4x + 1$ is a surjective function.

Solution 2.19. Let $f: U \subset \mathbb{R} \to M \subset \mathbb{R}$, suppose $\forall \ y \in M \subset \mathbb{R} \ \exists \ x \in U \subset \mathbb{R}$ such that $f(x) = y$. According to definition (Eq. 2.12) $f(x) = 4x + 1$ is surjective.

2.3.3. Bijection Type

Definition 2.18. A function that is **injective** and **surjective** is a **bijective** function.

Example 2.20. A function defined (Ex. 2.18–ii) is a bijective function. A function defined (Ex. 2.19) is a bijective function.

Remark 2.6. Any linear function $f: U \subset \mathbb{R} \to M \subset \mathbb{R}, ax + b$ where $a, b \in \mathbb{R}, a \neq 0$ is a bijection.

Definition 2.19. A map that is **injective** and **surjective** is a **bijective** map.

2.4. INVERSION

Definition 2.20. A **bijective** function f, whose **domain** is the set U and its **image** is the set M, has a unique **inverse function** f^{-1} where its **domain** is M and its **image** is U [30], with the property

$$f(x) = y \Leftarrow f^{-1}(y) = x.$$

Example 2.21. Define the inverse function of (i) $f: \mathbb{R} \to \mathbb{R}, x - 2$. (ii) $f: \mathbb{R}^+ \to \mathbb{R}^+, x^2$.

Solution 2.20. Since f is a bijective function, $\exists!\ f^{-1}: M \subset \mathbb{R}^+ \to U \subset \mathbb{R}^+, y + 2.$
(ii) Since f is a bijective function $\exists!\ f^{-1}: M \subset \mathbb{R}^+ \to U \subset \mathbb{R}^+, \sqrt{y}.$

Definition 2.21. A **bijective** map $T: U \subset \mathbb{R} \to M \subset \mathbb{R}, T_1(t)$ whose **domain** is the set U and its **image** is the set M, has an unique **inverse map** T^{-1} where its **domain** is M and its **image** is U with the property

$$T(t) = y \Leftarrow T^{-1}(y) = t.$$

2.5. COMPOSITION

Be the functions $f: U \subset \mathbb{R} \to M \subset \mathbb{R}$ and $g: U \subset \mathbb{R} \to M \subset \mathbb{R}$. The composition function is a **function** $f \circ g: f(g(x)): \mathbb{R} \to \mathbb{R}$. Note that the domain of function $f \circ g$ is the domain of function g and the image of $f \circ g$ is the image of function f. The composition function $f \circ g$ is different from the functions f or g (Fig. **2.7**). Figure taken from [29].

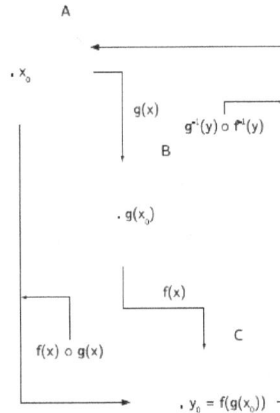

Fig. (2.7). The **function** $f(x) \circ g(x)$ goes from set A to C and the **function** $g^{-1} \circ f^{-1}$ goes from set C to A. Figure taken from [29].

Remark 2.7. The inverse composition function is $(f \circ g)^{-1} = g^{-1} \circ f^{-1}$, if there exists f^{-1} and g^{-1}.

Example 2.22. Be $f: \mathbb{R} \to \mathbb{R}$, $3x^2$ and $g: \mathbb{R} \to \mathbb{R}$, $\frac{x}{2}$. The composition function is the **function** $f \circ g: f(g(x)) = 3\left(\frac{x}{2}\right)^2.$

Example 2.23. Be $f: \mathbb{R} \rightarrow \mathbb{R}$, e^x and $g: \mathbb{R} \rightarrow \mathbb{R}$, $x - 1$. The composition function is the **function** $f \circ g: f(g(x)): \mathbb{R} \rightarrow \mathbb{R}$, $e^{(x-1)^2}$.

Example 2.24. Be $f: \mathbb{R} \rightarrow \mathbb{R}$, e^y and $g: \mathbb{R} \rightarrow \mathbb{R}$, $\log x$. The composition function is the **function** $f \circ g: f(g(x)): \mathbb{R} \rightarrow \mathbb{R}$, y.

2.6. ELEMENTARY FUNCTIONS AND MAPS

From now on, the different types of **functions** will be described and the equivalent **maps** will be illustrated with examples.

Definition 2.23. An **elementary** function (Def. 2.1) is a function whose terms are related with the operators: addition, substraction, multiplication, division, exponentation, or radicalization.

Depending on the nature of the terms integrating the elementary function, it can be defined as **algebraic** or **transcendental**.

Definition 2.24. An **algebraic** function is an **elementary** function whose terms satisfy a **polynomial expression** (Def. 2.25). Otherwise, the function will be a **transcendental** function [31].

Definition 2.25. A **polynomial equation** is an equation of the form $a_n x^n + a_{n-1} x^{n-1} + \cdots + a_2 x^2 + a_1 x + a_0 = 0$, where $n \in \mathbb{Z}^+$ and $a_i \in \mathbb{R}$ [32].

Example 2.25. Classify as algebraic or transcendental the following elementary functions. (i) $f(x) = \frac{1}{x}$. (ii) $f(x) = \sqrt{x}$. (iii) $f(x) = 2x^2 + 3x - 1$. (iv) e^x. (v) $f(x) = \cos x - \tan x$. (vi) $f(x) = x^{\frac{1}{x}}$. (vii) $x^e - \cos x$. (viii) 10^x.

Solution 2.21. Entries (i-iii) correspond to algebraic functions and entries (iv-viii) correspond to transcendental functions.

Definition 2.26. An **explicit** function is expressed in terms of the independent variable. An **implicit** function is expressed in terms of the **dependent** (Def. 2.1) and **independent** (Def. 2.1) variables [28].

2.6.1. Algebraic Functions and Maps

In this section, we will study the features of the algebraic functions or maps from the concepts previously explained. In the next sections, these functions will be explained from the concepts of differentiation and continuity.

The different types of **functions** will be described and the equivalent **maps** will be illustrated with examples.

2.6.2. Linear Type

Definition 2.27. A **linear** function is a function with a **polynomial degree** $n \leq 1$ $f: U \subset \mathbb{R} \rightarrow M \subset \mathbb{R}$, $ax^n + b$, where $a, b \in \mathbb{R}$, $n \in \mathbb{Z}^+$, $n \leq 1$ [33]. Its D_f is the set U, its I_f is the set M, its $\text{graph}(f) = \{(x, f(x)) \in \mathbb{R}^2 \mid x \in D_f, \text{and} f(x) = ax + b\}$ is a straight line (Fig. **2.8**).

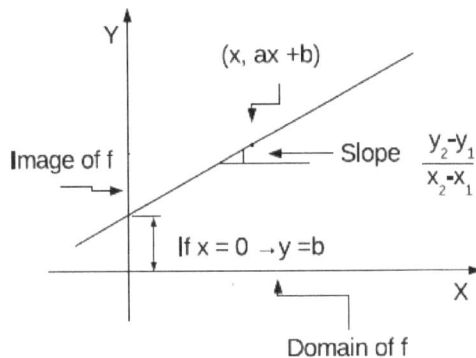

Fig. (2.8). Graph of $f(x) = ax + b$ and its components.

Example 2.26. Be the linear function $f: [0,2] \subset \mathbb{R} \rightarrow \mathbb{R}, 3x + 2$. (i) What is its D_f? (ii) What is its I_f? (iii) What is the value of a? (iv) What is the value of b? (v) At what point $x \in D_f$ has the function its maximum value? (vi) At what point $x \in D_f$ has the function its minimum value? (vii) What is its slope?

Solution 2.22. (i) $D_f = [x_1, x_2] = [0,2]$. (ii) $I_f = [f(x_1), f(x_2)] = [3(0) + 2, 3(2) + 2] = [2,8]$. (iii) $a = 3$. (iv) $b = 2$. (v) In $x_2 = 2 \Rightarrow f(x_2) = 8$, 8 the max is $[2,8] = I_f$. (vi) In $x_1 = 0 \Rightarrow f(x_1) = 2$, 2 the min is $[2,8] = I_f$. (vii) Be two points in the function f, $(x_1, y_1) = (0,2)$ and $(x_2, y_2) = (2,8)$, $m = \frac{y_2 - y_1}{x_2 - x_1} = \frac{8-2}{2-0} = \frac{6}{2} = 3$. The $angle = \tan^{-1}m = 71°$.

Example 2.27. Draw the graph ? for $1 - |x|$

Solution 2.23.

$$f(x) = \begin{cases} 1 - x & for & x > 0 \\ 1 - 0 & for & x = 0 \\ 1 - (-x) & for & x < 0 \end{cases} \qquad (2.14)$$

See (Fig. **2.9**).

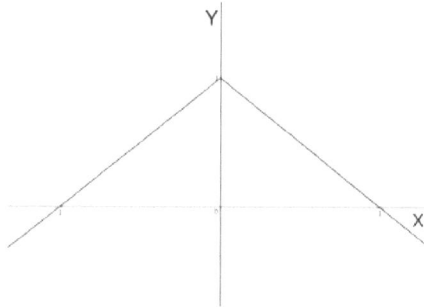

Fig. (2.9). Graph of $f(x) = 1 - |x|$.

Remark 2.8. The **existence** of the **inverse linear** function is conditioned to the **Implicit Function Theorem** (Sect. 5.7) and the definition of the **Inverse Function Theorem** (Sect. 5.8).

Example 2.28. Be the linear function $f(x) = 3x - 2$. (i) Determine the **inverse** function f^{-1} and its domain. (ii) Compute $f \circ f^{-1}$. (iii) Compute $f^{-1} \circ f$.

Solution 2.24. (i) $f^{-1}(y) = \frac{y+2}{3}$, its domain is $\mathbb{R}$. (ii) $f \circ f^{-1} = 3(\frac{y+2}{3}) - 2 = y$.
(iii) $f^{-1} \circ f = \frac{(3x-2)+2}{3} = x$.

Definition 2.28. The map (Def. 2.2.5) of a linear function (Def. 2.27) is $T: U \subset \mathbb{R} \to \mathbb{R}^2, (x, ax^n + b)$.

2.6.3. Quadratic Type

Definition 2.29. A **quadratic** function is a **polynomial** function with polynomial degree $n = 2$ $f: U \subset \mathbb{R} \to M \subset \mathbb{R}$, $ax^2 + bx + c$, where $a, b, c \in \mathbb{R}$, $n \in \mathbb{Z}^+$, $n = 2$?. Its D_f is the set U, its I_f is the set M, its $graph(f) = \{(x, f(x)) \in \mathbb{R}^2 \mid x \in D_f,, and f(x) = ax^2 + bx + c\}$ is a parabola (Fig. **2.10**).

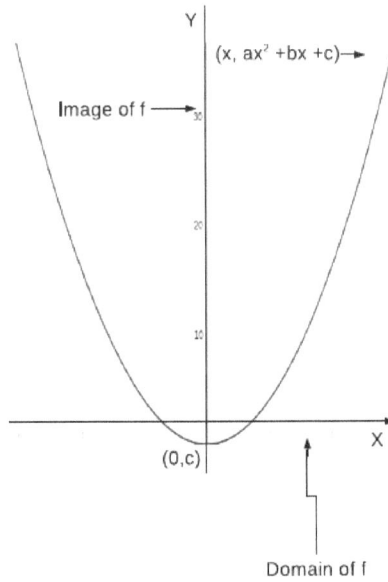

Fig. (2.10). Graph of $f(x) = ax^2 + bx + c$ and its components over the real oriented line.

Example 2.29. Be the quadratic function $f: [0,2] \subset \mathbb{R} \to \mathbb{R}, x^2 + 3x + 2$. (i) What is its D_f? (ii) What is its I_f? (iii) Find the points $x \in D_f$ such that $f(x) = 0$. (iv) Discuss the behavior of the graph.

Solution 2.25. (i) $D_f = [x_1, x_2] = [0,2]$. (ii) $I_f = [f(x_1), f(x_2)] = [0^2 + 3(0) + 2, 2^2 + 3(2) + 2] = [2,12]$. (iii) From $\frac{-b \pm \sqrt{b^2 - 4ac}}{2a} \Rightarrow x_1 = -1$ and $x_2 = -2$, then $\mathbf{6}\ x \in D_f$ such that $f(x) = 0$. (iv) It is a curve that opens upward from the point $(0,2)$ to the point $(2,12)$. The function f has a minimum value in $x = 0$ and a maximum value in $x = 2$. The slope of the curve changes at each point, from almost zero degrees at $(0, c)$, to almost $\frac{\pi}{2}$ radians at $(2,12)$.

Remark 2.9. The **existence** of the **inverse quadratic** function is conditioned to the **Implicit Function Theorem** (Sect. 5.7) and its definition to the **Inverse Function Theorem** (Sect. 5.8).

Example 2.30. Be the quadratic function $f(x) = 3x^2 + 1$. (i) Determine its **inverse** function f^{-1} and its domain. (ii) Compute $f \circ f^{-1}$. (iii) Compute $f^{-1} \circ f$.

Solution 2.26. (i) $f^{-1}(y) = \sqrt{\frac{y-1}{3}}$, its domain is $[1, \infty]$ (only the positive root was taken). (ii) $f \circ f^{-1} = 3(\sqrt{\frac{y-1}{3}})^2 + 1 = y$. (iii) $f^{-1} \circ f = \sqrt{\frac{(3x^2+1)-1}{3}} = x$.

Definition 2.30. The map (Def. 2.2.5) of a quadratic function (Def. 2.29) is $T: U \subset \mathbb{R} \to \mathbb{R}^2, (x, ax^2 + bx + c)$.

2.6.4. Rational Type

Definition 2.31. A **rational** function is a function that can be written as the ratio of two polynomial functions $f: U \subset \mathbb{R} \to M \subset \mathbb{R}, \frac{P(x)}{Q(x)}$, where $Q(x) \neq 0$?. Its D_f is the set of all the values of x whose denominator $Q(x)$ is not zero. Its I_f is the set M. Its graph$(f) = \{(x, f(x)) \in \mathbb{R}^2 \mid x \in D_f, \text{and} f(x) = \frac{P(x)}{Q(x)}, Q(x) \neq 0\}$ is the set of asymptote curves between singularities (Fig. **2.11**).

Remark 2.10. A **linear** function (Def. 6.2) such as $f(x) = c$ is a **rational** function, since its constants are polynomials [35].

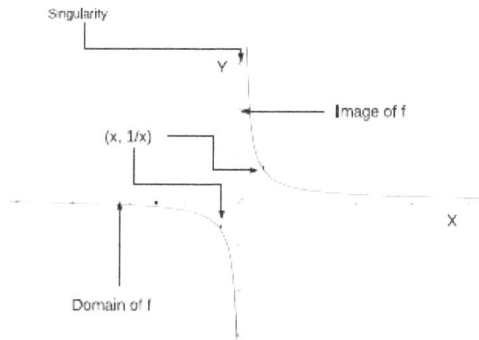

Fig. (2.11). Graph of the rational function $f(x) = \frac{1}{x}$, its singularity and components. Graph plotter [36].

Example 2.31. Be the rational function $f: \mathbb{R} \to \mathbb{R}, \frac{1}{x}$ (Fig. **2.11**). (i) What is its D_f? (ii) What is its I_f? (iii) Find the points $x \in D_f$ such that $f(x) = 0$. (iv) Discuss the behavior of the graph.

Solution 2.27. (i) $D_f = \mathbb{R} - \{0\}$. (ii) $I_f = \mathbb{R}$. (iii) ó $x \in D_f$ such that $f(x) = 0$. (iv) It has two asymptotic curves with one singularity in $x = 0$. The function f has neither a minimum value nor a maximum value. Observe that in this rational function a small positive input value yields a large positive output value; a complementary situation occurs with the negative values [37].

Remark 2.11. The **existence** of the **inverse rational** function is conditioned to the **Implicit Function Theorem** (Sect. 5.7) and the definition of the **Inverse Function Theorem** (Sect. 5.8).

Example 2.32. Be the rational function $f(x) = \frac{1}{x^2}$ defined in $\mathbb{R} - \{0\}$. (i) Determine its **inverse** function f^{-1} and its domain. (ii) Compute $f \circ f^{-1}$. (iii) Compute $f^{-1} \circ f$.

Solution 2.28. (i) $f^{-1}(y) = \frac{1}{\sqrt{y}}$, its domain is $(0, \infty)$. (ii) $f \circ f^{-1} = \frac{1}{(\frac{1}{\sqrt{y}})^2} = y$. (iii)

$$f^{-1} \circ f = \frac{1}{\sqrt{\frac{1}{x^2}}} = x.$$

Definition 2.32. The map (Def. 2.2.5) of a rational function (Def. 2.31) is $T: U \subset \mathbb{R} \to \mathbb{R}^2, (x, \frac{P(x)}{Q(x)})$.

2.6.5. Irrational Type

Definition 2.33. An **irrational** function is a function whose analytic expression is under the exponentation $f: U \subset \mathbb{R} \to M \subset \mathbb{R}, \sqrt[n]{g(x)}$, where $n \in \mathbb{N}$ and g are irrational functions ?. Its $D_f = D_g$. Its I_f is the set M. Its graph$(f) = \{(x, f(x) \in \mathbb{R}^2 \mid x \in D_g$ and $f(x) = \sqrt[n]{g(x)}, g: \mathbb{R} \to \mathbb{R}\}$ (Fig. **2.12**).

Example 2.33. Be the irrational function $f: \mathbb{R} \to \mathbb{R}, \sqrt{x}$ (Fig. **2.12**). (i) What is its D_f? (ii) What is its I_f? (iii) Find the points $x \in D_f$ such that $f(x) = 0$. (iv) Discuss the behavior of the graph.

Solution 2.29. (i) $D_f = D_g = \mathbb{R}^+$. (ii) $I_f = \mathbb{R}^+$. (iii) $x = 0$. (iv) The curve is located in the first quadrant and it is ascending [38].

Remark 2.12. The **existence** of the **inverse irrational** function is conditioned to the **Implicit Function Theorem** (Sect. 5.7) and the definition of the **Inverse Function Theorem** (Sect. 5.8).

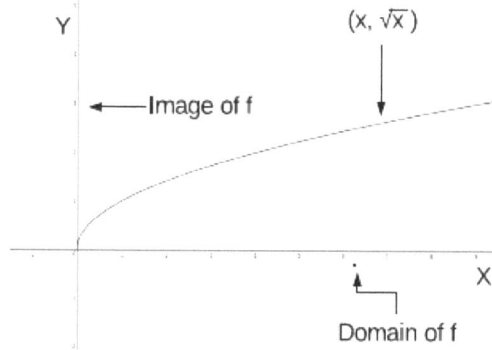

Fig. (2.12). Graph of the irrational function $f(x) = \sqrt{x}$ and components. Graph plotter ?.

Example 2.34. Be the irrational function $f(x) = \sqrt{x-1}$ defined in $(1, \infty)$. (i) Determine its **inverse** function f^{-1} and its domain. (ii) Compute $f \circ f^{-1}$. (iii) Compute $f^{-1} \circ f$.
Solution 2.30. (i) $f^{-1}(y) = y^2 + 1$, its domain is $\mathbb{R}$. (ii) $f \circ f^{-1} = \sqrt{(y^2+1)-1} = y$. (iii) $f^{-1} \circ f = (\sqrt{x-1})^2 + 1 = x$.

Definition 2.34. The map (Def. 2.2.5) of the irrational function (Def. 2.33) is $T: U \subset \mathbb{R} \to \mathbb{R}^2, (x, \sqrt[n]{g(x)})$.

2.6.6. Transcendental Functions and Maps

In this section, we will study the characteristics of **transcendental** functions from the concepts explained so far. In subsequent sections, these functions will be explained from the concepts of differentiation and continuity.

The different types of **functions** will be described and the corresponding **maps** will be illustrated with examples.

2.6.7. Sine Type

Definition 2.35. The **sine** function is a periodic function $f: [0,2\pi] \subset \mathbb{R} \to [-1,1] \subset \mathbb{R}$, $\sin x$. Its period is 2π, its $\text{graph}(f) = \{(x, f(x)) \in \mathbb{R}^2 \mid x \in D_f, \text{and} f(x) = \sin x\}$ (Fig. **2.13**).

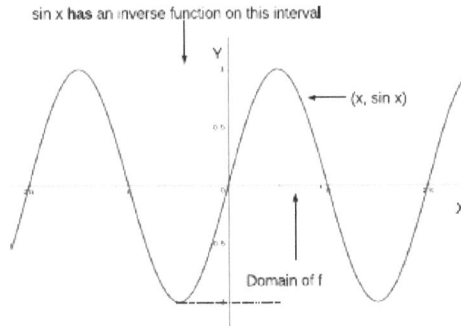

Fig. (2.13). Graph of the periodic trigonometric function $f(x) = \sin x$ and components, over the interval $[-2\pi, 2\pi]$. Graph plotter [36].

Example 2.35. Be the trigonometric function $f: [0,2\pi] \subset \mathbb{R} \to [-1,1] \subset \mathbb{R}, \sin x$ (Fig. **2.13**). (i) What is its D_f? (ii) What is its I_f? (iii) Find the points $x \in D_f$ such that $f(x) = 0$. (iv) Discuss the behavior of the graph.

Solution 2.31. (i) $D_f = [0,2\pi]$. (ii) $I_f = [-1,1]$. (iii) $x = 0, \pi$ and 2π. (iv) The image of f is positive in $[0, \pi]$ and negative in $[\pi, 2\pi]$, its period is 2π.

Definition 2.36. The **inverse** function (Def. 4) of $\sin x$ is $f^{-1}: [-1,1] \subset \mathbb{R} \to \left[-\frac{\pi}{2}, -\frac{\pi}{2}\right] \subset \mathbb{R}, \sin^{-1} x$ such that $f \circ f^{-1} = \sin(\sin^{-1} y) = y$, or $f^{-1} \circ f = \sin^{-1}(\sin x) = x$.

Remark 2.13. Note that the inverse function is **bijective** (Def. 2.3.3) in the domain $[-1,1]$ (Fig. **2.13**). The **existence** of the **inverse** function is conditioned to the **Implicit Function Theorem** (Sect. 5.7) and the definition of the **Inverse Function Theorem** (Sect. 5.8).

Definition 2.37. The map (Def. 2.2.5) to sine function (Def. 2.35) is $T: [0,2\pi] \subset \mathbb{R} \to \mathbb{R}^2, (x, \sin x)$.

2.6.8. Cosine Type

Definition 2.38. The **cosine** function is a periodic function $f: [0,2\pi] \subset \mathbb{R} \to [-1,1] \subset \mathbb{R}$, $\cos x$. Its period is 2π, its $\text{graph}(f) = \{(x, f(x)) \in \mathbb{R}^2 \mid x \in D_f,$ and $f(x) = \cos x\}$ (Fig. **2.14**).

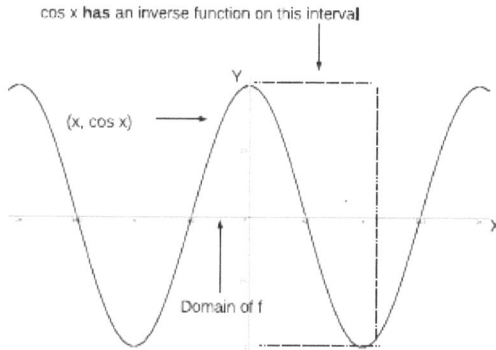

Fig. (2.14). Graph of the periodic trigonometric function $f(x) = \cos x$ and its components over the interval $[-2\pi, 2\pi]$. Graph plotter [36].

Example 2.36. Be the trigonometric function $f: [0,2\pi] \subset \mathbb{R} \to [-1,1] \subset \mathbb{R}, \cos x$ (Fig. **2.13**). (i) What is its D_f? (ii) What is its I_f? (iii) Find the points $x \in D_f$ such that $f(x) = 0$. (iv) Discuss the behavior of the graph.

Solution 2.32. (i) $D_f = [0,2\pi]$. (ii) $I_f = [-1,1]$. (iii) $x = \dfrac{\pi}{2}$ and $\dfrac{3\pi}{2}$. (iv) The image of f is positive in $[0, \dfrac{\pi}{2}] \cup [\dfrac{3\pi}{2}, 2\pi]$ and negative in $[\dfrac{\pi}{2}, \dfrac{3\pi}{2}]$, its period is 2π.

Definition 2.39. The **inverse** function (Def. 4) of $\cos x$ is $f^{-1}: [-1,1] \subset \mathbb{R} \to \left[0, \dfrac{\pi}{2}\right] \subset \mathbb{R}, \cos^{-1} x$ such that $f \circ f^{-1} = \cos(\cos^{-1} y) = y$, or $f^{-1} \circ f = \cos^{-1}(\cos x) = x$.

Remark 2.14. Note that the inverse function is **bijective** (Def. 2.3.3) in the domain $[-1,1]$ (Fig. **2.14**). The **existence** of the **inverse** function is conditioned to the **Implicit Function Theorem** (Sect. 5.7) and the definition of the **Inverse Function Theorem** (Sect. 5.8).

Definition 2.49. The map (Def. 2.2.5) to cosine function (Def. 2.38) is $T: [0,2\pi] \subset \mathbb{R} \to \mathbb{R}^2, (x, \cos x)$.

2.6.9. Tangent Type

Definition 2.41. The **tangent** function is a periodic function $f: [-\frac{\pi}{2}, \frac{\pi}{2}] \subset \mathbb{R} \to \subset \mathbb{R}$, $\tan x$. Its period is π, its $\text{graph}(f) = \{(x, f(x)) \in \mathbb{R}^2 \mid x \in D_f, \text{and} f(x) = \tan x\}$ (Fig. **2.15**).

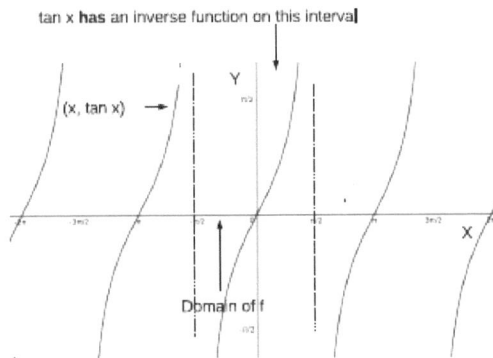

Fig. (2.15). Graph of the periodic trigonometric function $f(x) = \tan x$ and its components, over the interval $[-2\pi, 2\pi]$. Graph plotter [36].

Example 2.37. Be the trigonometric function $f: [-\frac{\pi}{2}, \frac{\pi}{2}] \subset \mathbb{R} \to \subset \mathbb{R}$, $\tan x$ (Fig. **2.14**). (i) What is its D_f? (ii) What is its I_f? (iii) Find the points $x \in D_f$ such that $f(x) = 0$. (iv) Discuss the behavior of the graph.

Solution 2.33. (i) $D_f = [-\frac{\pi}{2}, \frac{\pi}{2}]$. (ii) $I_f = \mathbb{R}$. (iii) $x = 0$. (iv) The curve is asymptotic to $x = -\frac{\pi}{2}$ and $x = -\frac{\pi}{2}$, its period is π.

Definition 2.42. The **inverse** function (Def. 4) of $\tan x$ is $f^{-1}: \mathbb{R} \to [-\frac{\pi}{2}, \frac{\pi}{2}] \subset \mathbb{R}$, $\tan^{-1} x$ such that $f \circ f^{-1} = \tan(\tan^{-1} y) = y$, or $f^{-1} \circ f = \tan^{-1}(\tan x) = x$.

Remark 2.15. Note that the inverse function is **bijective** (Def. 2.3.3) in the domain $\mathbb{R}$ (Fig. **2.15**). The **existence** of the **inverse** function is conditioned to the **Implicit Function Theorem** (Sect. 5.7) and the definition of the **Inverse Function Theorem** (Sect. 5.8).

Definition 2.43. The map (Def. 2.2.5) to tangent function (Def. 2.41) is $T: [-\frac{\pi}{2}, \frac{\pi}{2}] \subset \mathbb{R} \to \mathbb{R}^2, (x, \tan x)$.

2.6.10. Exponential Type

Definition 2.44. An **exponential** function is a function whose analytic expression is $f: U \subset \mathbb{R} \to M \subset \mathbb{R}^+$, e^x. Its $D_f = \mathbb{R}$, its I_f is the set M, and its $\text{graph}(f) = \{(x, f(x)) \in \mathbb{R}^2 \mid x \in \mathbb{R}, f(x) = e^x\}$ (Fig. **2.16**).

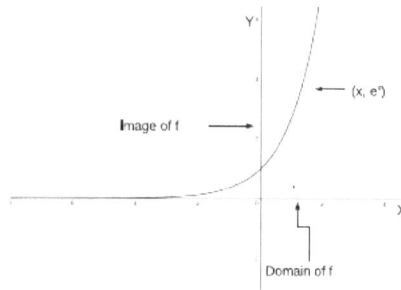

Fig. (2.16). Graph of the exponential function $f(x) = e^x$ and its components. Graph plotter [36].

Example 2.38. Be the exponential function $f: \mathbb{R} \to \mathbb{R}, e^x$ (Fig. **2.16**). (i) What is its D_f? (ii) What is its I_f? (iii) Find the points $x \in D_f$ such that $f(x) = 0$. (iv) Discuss the behavior of the graph.

Solution 2.34. (i) $D_f = U \subset \mathbb{R}$. (ii) $I_f = M \subset \mathbb{R}^+$. (iii) ó $x \in D_f$ such that $f(x) = 0$. (iv) The curve is asymptotic towards Quadrant II and ascending in Quadrant I. *Remark 2.16.* The **existence** of the **inverse** **exponential** function is conditioned to the **Implicit Function Theorem** (Sect. 5.7) and the definition of the **Inverse Function Theorem** (Sect. 5.8).

Definition 2.45. The map (Def. 2.2.5) to exponential function (Def. 2.44) is $T: U \subset \mathbb{R} \to \mathbb{R}^2, (x, e^x)$.

2.6.11. Logarithmic Type

Definition 2.46. A **logarithmic** function is a function whose analytic expression is $f: U \subset \mathbb{R}^+ \to M \subset \mathbb{R}$, $\ln x$. Its $D_f = \mathbb{R}^+$, its I_f is the set M, and its $\text{graph}(f) = \{(x, f(x)) \in \mathbb{R}^2 \mid x \in \mathbb{R}, f(x) = \ln x\}$ (Fig. **2.18**).

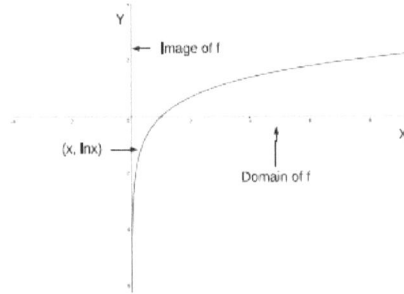

Fig. (2.18). Graph of the logarithmic function $f(x) = \ln x$ and components. Graph plotter [36].

Example 2.39. Be the exponential function $f(x) = e^x$. (i) Determine its **inverse** function f^{-1}. Determine its domain. (ii) Compute $f \circ f^{-1}$. (iii) Compute $f^{-1} \circ f$.

Solution 2.35. (i) $f^{-1}(y) = \ln y$. Its domain is $\mathbb{R}$. (ii) $f \circ f^{-1} = e^{(\ln y)} = y$. (iii) $f^{-1} \circ f = \ln(e^x) = x$.

Remark 2.17. The **existence** of the **inverse logarithmic** function is conditioned to the **Implicit Function Theorem** (Sect. 5.7) and the definition of the **Inverse Function Theorem** (Sect. 5.8).

Definition 2.47. The map (Def. 2.2.5) to logarithmic function (Def. 2.46) is $T: U \subset \mathbb{R} \to \mathbb{R}^2, (x, \ln x)$.

2.7. CASE STUDY: HAS $f(x) = x^2$ AN INVERSE?

Case 2.1. With the purpose of knowing if the function $f(x) = x^2$ has an inverse and what it is, we will begin by classifying the function.

We must bear in mind that when a function is stated without specifying its domain or image, it has to be assumed that both spaces act on the broader set. In this case, its domain and image is the set $\mathbb{R}$.

Note that this function is neither injective nor surjective, therefore, it is not a bijective function. (Ex. 2.16).

Let us look at the reason for this statement. Be two elements a and $-a$ from the domain of the function, when evaluating them in the function it yields $f(-a) = a^2 = f(a)$. This implies that f is not injective (Def. 2.3.1).

Note element $-a$ that belongs to the image of the function, does not come from any element in the domain *i.e.*, $\nexists x \in D_f$ such that $f(x) = -a$. This implies that f is not surjective (Def. 2.3.2).

These two results determine that f is not bijective (Def. 2.3.3). Why is it important to know that F is a bijective function?

Here, we will proof that if f is bijective then it has an inverse.

Theorem 2.1. *Let* $f: U \subset \mathbb{R} \to M \subset \mathbb{R}$ *is surjective, then* f *has an inverse.*

Proof. Let's define an inverse function $f^{-1}: M \subset \mathbb{R} \to U \subset \mathbb{R}$, now let's take element $m \in M$, if f is surjective there must exists an element $f(a) = b$ and $f^{-1}(b) = a$. Since f is injective, then $f^{-1} \circ f(a) f^{-1}(f(a) = f^{-1}(b) = a$ and $f \circ f^{-1}(b) f(f^{-1}(b) = f(a) = b$.

Note that we have only proven that if a function is bijective then it has an inverse and not the converse, that if a function has an inverse then it is bijective. We will not do that now, as we only want to know if a function has an inverse.

Now, we solve x from $f(x) = x^2$ and substitute $f(x)$ for y, then we find that $x = \pm\sqrt{y} \to g(x) = \pm\sqrt{y}$. $g(y)$ **is not** a function (Def. 2.1).
However, if we limit the domain and image of f to make it bijective, for example if the domain is $\mathbb{R}^+$ g limits $g(y) = +\sqrt{y}$, now it **is** a function, then $g(y) = f^{-1}(y)$. (Def. 2.33),

Let's look at the results graphically, in Fig. (**2.18**) the graph of f is a parabole, we can see the two functions separately $f^{-1}(x) \pm \sqrt{x}$. The graph of the two possible inverses is not a function; however, when the domain of f is limited to make it bijective, this inconsistency is solved.

Let's check the usefulness of a map (Def. 2.2.5) in this analysis. The map $T(t) = (t, t^2), t \in [-2.1, 2.1]$ is equivalent to $f: [-2.1, 2.1] \subset \mathbb{R} \to \mathbb{R}, x^2$. The advantage of map functions is that they **do not** have restrictions (Def. 2.1).

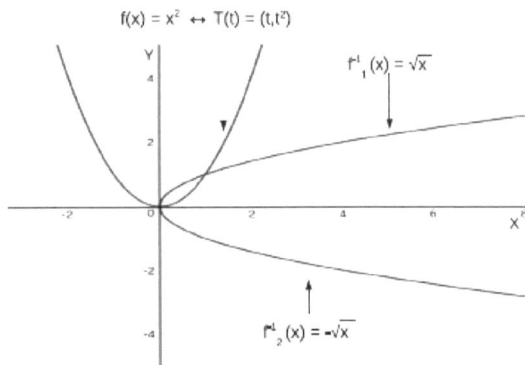

Fig. (2.18). Graph of the function $f(x) = x^2$ and its possible inverse functions $f^{-1}(x)$. Graph plotter [36].

2.8. CASE STUDY: MAXIMUM AREA TRIANGLE

Case 2.2. The objective is to determine, from all the triangles rectangles of hypotenuse 4, the one with the maximum area. As you can see in Fig. (**2.19**) there are infinite right triangles whose hypotenuse is 4, this means there are infinite areas.

We know that the area A of a right triangle depends on the variables x and y (Eq. 2.15), where x, y are the sides of the right triangle. In addition, the hypotenuse h meets (Eq. 2.16).

$$A(x, y) = \frac{xy}{2} \tag{2.15}$$
$$h^2 = x^2 + y^2$$

$$16 = x^2 + y^2 \tag{2.16}$$

$$y = \sqrt{16 - x^2}$$

If we take both equations, the area of the right triangle is a function $A(x)$ in terms of the variable x (Eq. 2.17).

$$A(x) = \frac{x\sqrt{16 - x^2}}{2} = \frac{\sqrt{16x^2 - x^4}}{2} \tag{2.17}$$

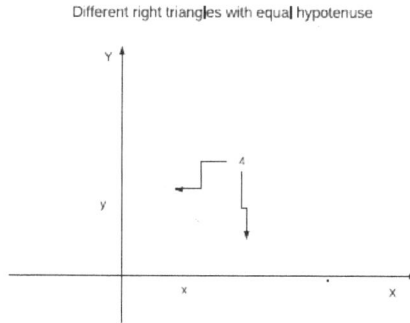

Different right triangles with equal hypotenuse

Fig. (2.19). Graph of right triangle. Graph plotter [36].

Looking at (Eq. 2.17), we notice that variable x must be positive since it represents one side of the triangle, it cannot be zero because the area of the right triangle would be zero.

When graphing function $A(x) = \frac{\sqrt{16x^2 - x^4}}{2}$ (Fig. **2.20**), we can see that $A(x)$ has a maximum in $\sqrt{8}$.

Remark 2.18. We will see this case again in Chap. 5, using the derivative of a function operator.

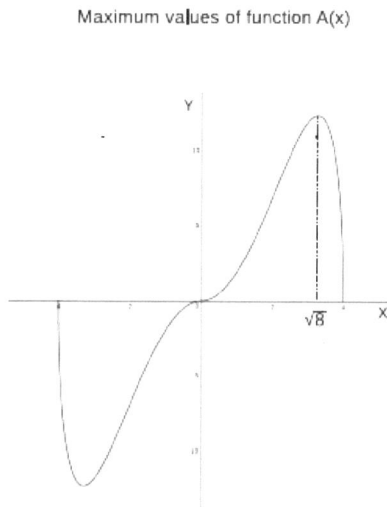

Maximum values of function A(x)

Fig. (2.20). Graph of the function $A(x) = \frac{\sqrt{16x^2 - x^4}}{2}$ and the point $\sqrt{8}$. Graph plotter [36].

2.9. CASE STUDY: EXPECTED PROFIT LABORATORIES

Case 2.3. A medication supplier sells a box of antibiotic with 30 pills for 16 Euros. Calculate the following:

- The function g calculates the number of pills sold in function of the number of boxes.
- The function f calculates the profit of the supplier in function of the boxes sold.

- The mÃƒÂ-nimum number of pills that have to be sold to obtain a profit of 4100 Euros.

- The profit of 400 pills sold

 Variable x will be the number of boxes sold and function $g(x)$ the number of pills sold. Since the box has 30 pills the function will be (Eq. 2.18)

$$g(x) = 30x \qquad (2.18)$$

The graph for (Eq. 2.18) is (Fig. **2.21**).

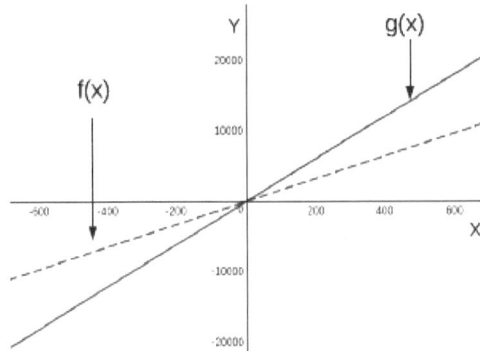

Fig. (2.21). Graph of the functions $g(x) = 30x$ and $f(x) = 16x$. Graph plotter [36].

For function f, x will be the number of boxes sold and $f(x)$ will be the total profit. Since each box costs 16 Euros, the function will be (Eq. 2.19)

$$f(x) = 16x \qquad (2.19)$$

The graph of (Eq. 2.19) is (Fig. **2.21**).

The profit is 4100 Euros for x number of boxes, therefore, $f(x) = 4100$. In other words, $16x = 4100 \Rightarrow x = \frac{4100}{16} = 256.25$, *i.e.* 257 boxes have to be sold to get a profit of 4100 Euros.

Since each box has 30 pills, the number of pills to be sold is $30 \times 257 = 7710$.

Finally, let's determine the profit if 400 pills are sold. Since each box has 30 pills, 400 will be $\frac{400}{30} = 13$ boxes, the profit will be $f(18) = 16 \times 18 = 288$ Euros.

Thus, if 400 are sold the profit will be 288 Euros.

2.10. EXERCISES

Exercise 2.1. Let the non-numeric function $f: \mathbb{A} \to \mathbb{B}$ allocates the age "80" to "old person" or "20" to "young person". (i) Obtain $f(x)$ for the elements $x = \{$oldperson, young person$\}$. (ii) Is f a function?

Exercise 2.2. Let the function $f: \mathbb{R} \Rightarrow \mathbb{R}, \sqrt[3]{x+3}$ and $f: \mathbb{R} \Rightarrow \mathbb{R}, \frac{1}{x}$. Obtain (i) $(f + g)(x)$. (ii) $(f - g)(x)$. (iii) $(fg)(x)$.

Exercise 2.3. Be $f(x) = \sqrt[5]{|x|}$. (i) Describe this function using an equivalent notation. (ii) What is the domain of f? (iii) What is the image of f?. (iv) What is its graph?

Exercise 2.4. (i) Is the function $f: \mathbb{R} \to \mathbb{R}, \frac{1}{x+3}$ an injective function? (ii) What is its graph?

Exercise 2.5. Is true that $f^{-1}(x) = \frac{1}{f(x)}$? Provide an example.

Exercise 2.6. Be $f: \mathbb{R} \to \mathbb{R}, x - 2$ and $g: \mathbb{R} \to \mathbb{R}, \ln x$. (i) Determine the composition function $f \circ g$. (ii) Determine the composition function $g \circ f$.

Exercise 2.7. Be $f: \mathbb{R} \to \mathbb{R}, x - 2$. (i) Determine the inverse function of f. (ii) Determine the composition function $f \circ f^{-1}$. (iii) Determine the composition function $f^{-1} \circ f$. (iv) Discuss the results of (ii) and (iii).

Exercise 2.8. Be the map $T_1: [0,2] \subset \mathbb{R} \to \mathbb{R}, (\frac{x}{2})$ and the map $T_2: [0,1] \subset \mathbb{R} \to \mathbb{R}^2, (t,t)$. (i) Compute $T_2 \circ T_1$. (ii) Graphically describe the composition (i).

Exercise 2.9. Be the function $f_1(x) = \sin 2x$ and the function $f_2(x) = \cos x^2$. (i) Compute $f_2 \circ f_1$. (ii) Compute $f_1 \circ f_2$.

Exercise 2.10. Does the graph $x^2 + y^2 = 1$ come from a function? (i) Explain. (ii) Determine an equivalent mapping T.

CHAPTER 3

Limits

Abstarct: This chapter introduces the limit operator of a function and a map, its properties and procedure. This operator has an important role in the derivative and integral operators.

Keywords: Existence of the limits of a function and a map, Limit of logarithmic function, Limit of logarithmic map, Limit of a function, Limit of a map, Limit of exponential function, Limit of exponential map, Limit of polynomial function, Limit of polynomial map, Limit of trigonometric function, Limit of trigonometric map, One-sided limit of a function and a map, Properties of the limit of a function and a map, Two-sided limits of a function and a map.

3.1. FUNCTIONS

3.1.1. Limit of a Function

The **Limit** of a function f is used in calculus and it refers to the proximity between a value x and a point x_0. For example, if a function f has an x limit at a point x_0 (Fig. **3.1**), it means that the value of f can be as close to x as desired, with points close enough to x_0 but not equal to x_0. ?. It is said that a function f has a limit L at the point $x = x_0$, if it is possible to indefinitely approach x to x_0 and it remains different from x_0 [40].

Definition 3.1. Let **function** f be defined with an interval in the neighbourhood of x_0, where $(f(x_0)$ does not necessarily have to be defined. It is said that the limit of f as x approaches or converges to x_0 is L [41],

$$\lim_{x \to x_0} f(x) = L$$

if for every $\varepsilon > 0$ there exists $\delta > 0$, such that for all x

$$0 < |x - x_0| < \delta \Rightarrow |f(x) - L| < \varepsilon.$$

Carlos Polanco
All rights reserved-© 2020 Bentham Science Publishers

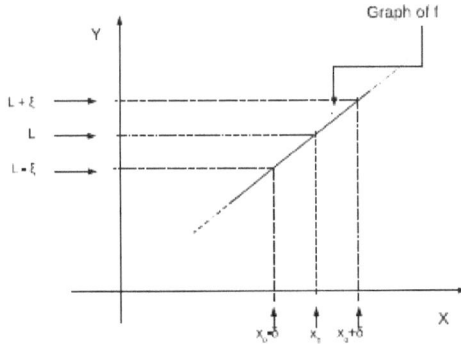

Fig. (3.1). Graphic description of limit L in function f.

Remark 3.1. If the limit of f at x_0 does not exist, it is said that f diverges at x_0.

Theorem 3.1. Uniqueness of Limits. Let the function $f: \mathbb{R} \to \mathbb{R}$. If $\lim_{x \to x_0} f(x) = L_1$ and $\lim_{x \to x_0} f(x) = L_2$, then $L_1 = L_2$.

Proof. Suppose that $L_1 \neq L_2$, then $\lim_{x \to x_0} f(x) = L_1 \Leftrightarrow |f(x) - L_1| < \varepsilon_1 = \frac{\varepsilon}{2}$ and $\lim_{x \to x_0} f(x) = L_2 \Leftrightarrow |f(x) - L_2| < \varepsilon_2 = \frac{\varepsilon}{2}$.

$$|L_1 - L_2| = |L - f(x) + f(x) - M| \leq |L_1 - f(x)| + |f(x) - L_2| < \varepsilon_1 + \varepsilon_2 = \varepsilon \quad \textbf{(3.1)}$$

But $\varepsilon > 0$, then $L_1 \neq L_2$ contradiction. So $L_1 = L_2$.

Example 3.1. Demonstrate using (ε/δ) $\lim_{x \to 2} (2x + 3) = 7$.

Proof. Show that if $|x - x_0| < \delta$, then $|f(x) - L| < \varepsilon$.

If $|x - 2| < \delta$, then $|(2x + 3) - 7| < \varepsilon$. Let $\delta = \frac{\varepsilon}{2}$ (Rmk. 3.1). If $|x - 2| < \delta = \frac{\varepsilon}{2} \Rightarrow 2|x - 2| < \varepsilon \Rightarrow |2x - 4| = |(2x + 3) - 7| < \varepsilon$.

Remark 3.2. $|(2x + 3) - 7| < \varepsilon \Rightarrow |2x - 4| < \varepsilon \Rightarrow 2|x - 2| < \varepsilon \Rightarrow |x - 2| < \frac{\varepsilon}{2}$.

Example 3.2. Demonstrate using (ε/δ) $\lim_{x \to 0} x^2 = 0$.

Proof. Show that if $|x - x_0| < \delta$, then $|f(x) - L| < \varepsilon$.

If $|x - 0| < \delta$, then $|(x^2) - 0| < \varepsilon$. Let $\delta = \sqrt{\varepsilon}$ (Rmk. 3.3). If $|x - 0| < \delta = \sqrt{\varepsilon} \Rightarrow |x - 0|^2 < \varepsilon \Rightarrow |x^2 - 0| = |(x^2) - 0| < \varepsilon$.

Remark 3.3. $|(x^2) - 0| < \varepsilon \Rightarrow |x^2| < \varepsilon \Rightarrow |x|^2 < \varepsilon \Rightarrow |x| < \sqrt{\varepsilon}$.

Example 3.3. Demonstrate using (ε/δ) $\lim_{x \to 0} x = 3$.

Proof. Show that if $|x - x_0| < \delta$, then $|f(x) - L| < \varepsilon$.

If $|x - 0| < \delta$, then $|(x) - 3| < \varepsilon$. From (Rmk. 3.3) $\delta = \varepsilon + 3$, δ should be greater than 3; so this limit does **not** exist. See (Fig. **3.2**). $|(x) - 3| < \varepsilon \Rightarrow |x| < \varepsilon + 3$.

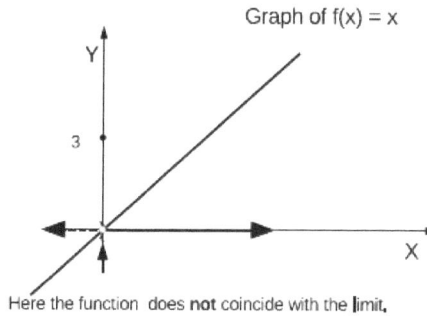

Fig. (3.2). Graphic description of $\lim_{x \to 0} x = 3$.

This definition is particularly used only in unusual situations. For many applications, it is highly recommended to use the following properties to answer straightforward questions about limits.

3.1.2. Properties of Limit of Functions

Let the functions $f: \mathbb{R} \to \mathbb{R}$ and $g: \mathbb{R} \to \mathbb{R}$, if $\lim_{x \to x_0} f(x) = L$, $\lim_{x \to x_0} g(x) = M$ and $\alpha \in \mathbb{R}$ then:

Property 1. $\alpha \lim_{x \to x_0} f(x) = \lim_{x \to x_0} \alpha f(x) = \alpha L$ (constant).
Property 2. $\lim_{x \to x_0} f(x) + g(x) = \lim_{x \to x_0} f(x) + \lim_{x \to x_0} g(x) = L + M$ (addition).
Property 3. $\lim_{x \to x_0} f(x) - g(x) = \lim_{x \to x_0} f(x) - \lim_{x \to x_0} g(x) = L - M$ (substraction).
Property 4. $\lim_{x \to x_0} f(x) \times g(x) = \lim_{x \to x_0} f(x) \times \lim_{x \to x_0} g(x) = L \times M$ (multiplication).

Property 5. If $M \neq 0$, $\lim_{x \to x_0} \frac{f(x)}{g(x)} = \frac{\lim_{x \to x_0} f(x)}{\lim_{x \to x_0} g(x)} = \frac{L}{M}$ (division).

3.1.3. Existence of One-sided Limits of a Function

Definition 3.2. Let f be a function defined in an interval $(-\infty, x_0)$, except at point x_0. The limit of function f on x_0 exists and it is equal to L if:

$$\lim_{x \to x_0^-} f(x) = L \in \mathbb{R}$$

$L \in \mathbb{R}$ is a **left-hand limit of** f **at** x_0.

Remark 3.5. It is important to mention that L must belong to $\mathbb{R}$, the symbols "∞" or "$-\infty$" **do not** represent numbers. They symbolize an increment or decrement without boundary. If the **limit** is "∞" or "$-\infty$", then it **does not** exist.

Example 3.4. Let the function $f : (\infty, -1] \subset \mathbb{R} \to \mathbb{R}, 2 - x$. Where is $\lim_{x \to -1^-} 2 - x$?

Solution 3.1. Since $\lim_{x \to -1^-} 2 = 2$ and $\lim_{x \to -1^-} x = 1$, then $\lim_{x \to -1^-} 2 - x = \lim_{x \to -1^-} 2 - \lim_{x \to -1^-} x = 2 - 1 = 1$.

Example 3.5. Let the function $f : (\infty, 1) \subset \mathbb{R} \to \mathbb{R}, x^2$. Where is $\lim_{x \to 1^-} x^2$?

Solution 3.2. Since $\lim_{x \to 1^-} x = 1$, then $\lim_{x \to 1^-} x^2 = [\lim_{x \to 1^-} x][\lim_{x \to 1^-} x] = (1)(1) = 1$.

Example 3.6. Let the function $f : (\infty, 1) \subset \mathbb{R} \to \mathbb{R}, \frac{x^2 + 2x - 3}{|x-1|}$. Where is $\lim_{x \to 1^-} f(x)$?

Solution 3.3. Since $x - 1 < 0$, then $|x - 1| = -(x - 1)$; so $\lim_{x \to 1^-} \frac{(x-1)(x+3)}{-(x-1)} = -x - 3 = -4$.

Definition 3.3. Let f be a function defined at the interval (x_0, ∞), except at point x_0; the limit of function f on x_0 exists and it is equal to L

$$\lim_{x \to x_0^+} f(x) = L \in \mathbb{R}$$

$L \in \mathbb{R}$ is a **right-hand limit of** f **at** x_0.

Remark 3.6. It is important to mention that L must belong to $\mathbb{R}$, the symbols "∞" or "$-\infty$" **do not** represent numbers. They symbolize an increment or decrement without boundary; if the **limit** is "∞" or "$-\infty$", then it **does not** exist.

Example 3.7. Let the function $f:[1,\infty) \subset \mathbb{R} \to \mathbb{R}, 3x^3 - 1$. Where is $\lim_{x \to 1^+} 3x^3 - 1$?

Solution 3.4. The $\lim_{x \to 1^+} 3 = 3$, $\lim_{x \to 1^+} x^3 = 1$ and $\lim_{x \to 1^+} 1 = 1$ exist; so the $\lim_{x \to 1^+} 3x^3 - 1 = 3[\lim_{x \to 1^+} x]^3 - \lim_{x \to 1^+} 1 = 3(1)^3 - 1 = 2$.

Example 3.8. Let the function $f:[0,\infty) \subset \mathbb{R} \to \mathbb{R}, \frac{1}{x}$. (i) Where is $\lim_{x \to 0^+} \frac{1}{x}$? (ii) What is the graph of f?

Solution 3.5. (i) The $\lim_{x \to 0^+} \frac{1}{x} = +\infty$, then the $\lim_{x \to 0^+} \frac{1}{x}$ **does not** exist. (Rmk. 3.6). (ii) See (Fig. **3.3**)

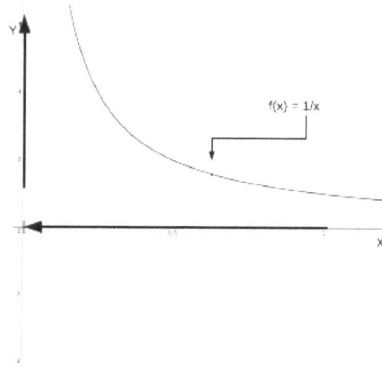

Fig. (3.3). The graph of $\frac{1}{x}$ in the interval $[0,\infty)$.

3.1.4. Existence of Two-sided Limits of a Function

Definition 3.4. From (Def. 3.2, 3.3) of a function defined at an interval, except at point x_0, we gather that the limit of the function f on x_0 exists and it is equal to $L \in \mathbb{R}$ if:

$$\lim_{x \to x_0^-} f(x) = L = \lim_{x \to x_0^+} f(x)$$

then,

$$\lim_{x \to x_0} f(x) = L$$

$L \in \mathbb{R}$ is a **two-sided limit of** f **at** x_0.

Remark 3.7. It is important to mention that L must belong to $\mathbb{R}$, the symbols "∞" or "−∞" **do not** represent numbers. They symbolize an increment or decrement without boundary. If the **limit** is "∞" or "−∞" then it **does not** exist.

Example 3.9. Let the function $f: \mathbb{R} \to \mathbb{R}, \frac{1}{x^2}$. Where is $\lim_{x \to 0} \frac{1}{x^2}$?

Solution 3.6. The $\lim_{x \to 0^-} \frac{1}{x^2} = +\infty$, $\lim_{x \to 0^+} \frac{1}{x^2} = +\infty$ are equal, but they are not numbers.

So $\lim_{x \to 0} \frac{1}{x^2}$ **does not** exist. (Rmk. 3.6).

Example 3.10. Let the function $f: \mathbb{R} \to \mathbb{R}, \sin\frac{\pi}{2x}$. Where is $\lim_{x \to 0} \sin\frac{\pi}{2x}$?

Solution 3.7. The values of I_f fluctuate with the pattern $1, 0, -1$ (Table **3.1**) and its fluctuation (Fig. **3.4**) increases as the interval is reduced to 0 (Fig. **3.5**). For this reason $\lim_{x \to 0} \sin\frac{\pi}{2x}$ **does not** exist.

Table 1. Evaluation of $f(x) = \sin\frac{\pi}{2x}$.

x	1	1/2	1/3	1/4	1/5	1/6	1/7	1/8
$f(x)$	1	0	−1	1	0	−1	1	0

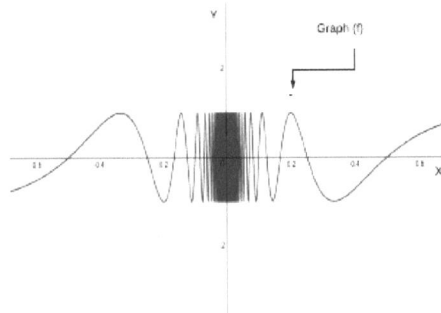

Fig. (3.4). Graph of $\sin\frac{\pi}{2x}$ in the interval $[-1,1]$.

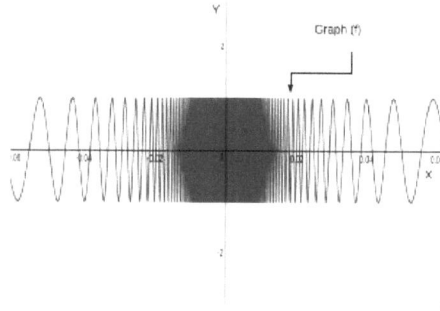

Fig. (3.5). Graph of $\frac{\pi}{2x}$ in the interval $[-0.1, 0.1]$.

3.1.5. Important Quote on Limit of a Function

The limit of a function f **is not** resolved by evaluating the function at point x_0. Note that the $\lim_{x \to 2} x^2 = \lim_{x \to 2^-} x^2 = 4$, but $f(x_0) = f(2)$ is not defined. In this case it is not possible to evaluate it.

Remark 3.8. Determining the limit of a function must always be carried out looking for the tendency of the boundary near the point x_0, *i.e.*, there is no general procedure to obtain it, each case is different.

3.2. MAPS

3.2.1. Limit of a Map

This section describes the limit of a map with some particular cases that form them and their graphical representation. This topic will be reviewed in Chap. 5-7. The **Limit** of a map T is used in Calculus and it refers to the proximity between a value t and a point t_0 in the domain of the map T. For example, if a map T has a t limit at a point t_0, it means that the value of T can be as close to t as desired, with points close enough to t_0 but not equal to t_0. [39]. It is said that a map T has a limit L at the point $t = t_0$, if it is possible to indefinitely approach t to t_0 and it remains different from t_0 [40].

Definition 3.5. Let **map** T be defined with an interval in the neighbourhood of t_0, where $(T(t_0)$ does not necessarily have to be defined. It is said that the limit of T as t approaches or converges to t_0 is L [41],

$$\lim_{t \to t_0} T(t) = \lim_{t \to t_0} T_1(t) = L$$

if for every $\varepsilon > 0$ there exists $\delta > 0$, such that for all t

$$0 < |t - t_0| < \delta \Rightarrow |T_1(t) - L| < \varepsilon.$$

Remark 3.9. If the limit of T at t_0 does not exist, it is said that T diverges at t_0.

Theorem 3.2. Uniqueness of Limits. Let the map $T: \mathbb{R} \to \mathbb{R}$. If $\lim_{t \to t_0} T(t) = L_1$ and $\lim_{t \to t_0} T(t) = L_2$, then $L_1 = L_2$.

Proof. Suppose that $L_1 \neq L_2$, then $\lim_{t \to t_0} T(x) = L_1 \Leftrightarrow |T(t) - L_1| < \varepsilon_1 = \frac{\varepsilon}{2}$ and $\lim_{t \to t_0} T(t) = L_2 \Leftrightarrow |T(t) - L_2| < \varepsilon_2 = \frac{\varepsilon}{2}$.

$$|L_1 - L_2| = |L - T(t) + T(t) - M| \leq |L_1 - T(t)| + |T(t) - L_2| < \varepsilon_1 + \varepsilon_2 = \varepsilon \qquad (3.2)$$

But $\varepsilon > 0$, then $L_1 \neq L_2$ contradiction. So $L_1 = L_2$.

Example 3.11. Demonstrate using (ε/δ) $\lim_{t \to 2} (2t + 3) = 7$.

Proof. Show that if $|t - t_0| < \delta$, then $|T(t) - L| < \varepsilon$.

If $|t - 2| < \delta$, then $|(2t + 3) - 7| < \varepsilon$. Let $\delta = \frac{\varepsilon}{2}$ (Rmk. 3.1). If $|t - 2| < \delta = \frac{\varepsilon}{2} \Rightarrow 2|t - 2| < \varepsilon \Rightarrow |2t - 4| = |(2t + 3) - 7| < \varepsilon$.

Remark 3.10. $|(tx + 3) - 7| < \varepsilon \Rightarrow |tx - 4| < \varepsilon \Rightarrow 2|t - 2| < \varepsilon \Rightarrow |t - 2| < \frac{\varepsilon}{2}$.

3.2.2. Properties of Limit of Maps

Let the maps $T_1: \mathbb{R} \to \mathbb{R}$ and $T_2: \mathbb{R} \to \mathbb{R}$, if $\lim_{t \to t_0} T_1(t) = L$, $\lim_{t \to t_0} T_2(x) = M$ and $\alpha \in \mathbb{R}$ then:

Property 1. $\alpha \lim_{t \to t_0} T_1(t) = \lim_{t \to t_0} \alpha T_1(t) = \alpha L$ (constant).
Property 2. $\lim_{t \to t_0} T_1(t) + T_2(t) = \lim_{t \to t_0} T_1(t) + \lim_{t \to t_0} T_2(t) = L + M$ (addition).
Property 3. $\lim_{t \to t_0} T_1(t) - T_2(t) = \lim_{t \to t_0} T_1(t) - \lim_{t \to t_0} T_2(t) = L - M$ (substraction).
Property 4. $\lim_{t \to t_0} T_1(t) \times T_2(t) = \lim_{t \to t_0} T_1(t) \times \lim_{t \to t_0} T_2(t) = L \times M$ (multiplication).

Property 5. If $M \neq 0$, $\lim_{t \to t_0} \dfrac{T_1(t)}{T_2(t)} = \dfrac{\lim_{t \to t_0} T_1(t)}{\lim_{t \to t_0} T_2(t)} = \dfrac{L}{M}$ (division).

3.2.3. Existence of One-sided Limits of a Map

Definition 3.6. Let f be a function defined in an interval $(-\infty, 0)$, except at point x_0. The limit of function f on x_0 exists and it is equal to L if:

$$\lim_{x \to x_0^-} f(x) = L \in \mathbb{R}$$

$L \in \mathbb{R}$ is a **left-hand limit of f** at x_0.

Remark 3.11. It is important to mention that L must belong to $\mathbb{R}$, the symbols "∞" or "$-\infty$" **do not** represent numbers. They symbolize an increment or decrement without boundary. If the **limit** is "∞" or "$-\infty$", then it **does not** exist.

Example 3.12. Let the function $f : (\infty, -1] \subset \mathbb{R} \to \mathbb{R}, 2 - x$. Where is $\lim_{x \to -1^-} 2 - x$?

Solution 3.8. Since $\lim_{x \to -1^-} 2 = 2$ and $\lim_{x \to -1^-} x = 1$, then $\lim_{x \to -1^-} 2 - x = \lim_{x \to -1^-} 2 - \lim_{x \to -1^-} x = 2 - 1 = 1$.

Example 3.13. Let the function $f : (\infty, 1) \subset \mathbb{R} \to \mathbb{R}, x^2$. Where is $\lim_{x \to 1^-} x^2$?

Solution 3.9. Since $\lim_{x \to 1^-} x = 1$, then $\lim_{x \to 1^-} x^2 = [\lim_{x \to 1^-} x][\lim_{x \to 1^-} x] = (1)(1) = 1$.

Example 3.14. Let the function $f : (\infty, 1) \subset \mathbb{R} \to \mathbb{R}, \dfrac{x^2 + 2x - 3}{|x - 1|}$. What is $\lim_{x \to 1^-} f(x)$?

Solution 3.10. Since $x - 1 < 0$, then $|x - 1| = -(x - 1)$; so $\lim_{x \to 1^-} \dfrac{(x-1)(x+3)}{-(x-1)} = -x - 3 = -4$.

Definition 3.7. Let f be a function defined at the interval (x_0, ∞), except at point x_0; the limit of function f on x_0 exists and it is equal to L

$$\lim_{x \to x_0^+} f(x) = L \in \mathbb{R}$$

$L \in \mathbb{R}$ is a **right-hand limit of f** at x_0.

Remark 3.12. It is important to mention that L must belong to $\mathbb{R}$, the symbols "∞" or "$-\infty$" **do not** represent numbers. They symbolize an increment or decrement without boundary; if the **limit** is "∞" or "$-\infty$", then it **does not** exist.

Example 3.15. Let the function $f:[1,\infty) \subset \mathbb{R} \to \mathbb{R}, 3x^3 - 1$. Where is $\lim_{x\to 1^+} 3x^3 - 1$?

Solution 3.11. The $\lim_{x\to 1^+} 3 = 3$, $\lim_{x\to 1^+} x^3 = 1$ and $\lim_{x\to 1^+} 1 = 1$ exist; so the $\lim_{x\to 1^+} 3x^3 - 1 = 3[\lim_{x\to 1^+} x]^3 - \lim_{x\to 1^+} 1 = 3(1)^3 - 1 = 2$.

Example 3.16. Let the function $f:[0,\infty) \subset \mathbb{R} \to \mathbb{R}, \frac{1}{x}$. (i) Where is $\lim_{x\to 0^+} \frac{1}{x}$? (ii) What is the graph of f?

Solution 3.12. (i) The $\lim_{x\to 0^+} \frac{1}{x} = +\infty$, then the $\lim_{x\to 0^+} \frac{1}{x}$ **does not** exist. (Rmk. 3.12). (ii) See (Fig. **3.3**).

3.2.4. Existence of Two-sided Limits of a Map

Definition 3.8. From (Def. 3.6, 3.7) of a function defined at an interval, except at point x_0, we know that the limit of the function f on x_0 exists and it is equal to $L \in \mathbb{R}$ if:

$$\lim_{x\to x_0^-} f(x) = L = \lim_{x\to x_0^+} f(x)$$

then,

$$\lim_{x\to x_0} f(x) = L$$

$L \in \mathbb{R}$ is a **two-sided limit of f at x_0.**

Remark 3.13. It is important to mention that L must belong to $\mathbb{R}$, the symbols "∞" or "$-\infty$" **do not** represent numbers. They symbolize an increment or decrement without boundary. If the **limit** is "∞" or "$-\infty$" then it **does not** exist.

Example 3.17. Let the function $f:\mathbb{R} \to \mathbb{R}, \frac{1}{x^2}$. Where is $\lim_{x\to 0} \frac{1}{x^2}$?

Solution 3.13. The $\lim_{x\to 0^-} \frac{1}{x^2} = +\infty$, $\lim_{x\to 0^+} \frac{1}{x^2} = +\infty$ are equal, but they are not numbers. So $\lim_{x\to 0} \frac{1}{x^2}$ **does not** exist. (Rmk. 3.6).

Example 3.18. Let the function f: $\mathbb{R} \to \mathbb{R}$, $\sin\frac{\pi}{2x}$. Where is $\lim_{x\to 0}\sin\frac{\pi}{2x}$?

Solution 3.14. The values of I_f fluctuate with the pattern $1, 0, -1$ (Fig. **6.3**) and its fluctuation (Fig. **3.4**) increases as the interval is reduced to 0 (Fig. **3.5**). For this reason $\lim_{x\to 0}\sin\frac{\pi}{2x}$ **does not** exist.

3.3. SOME BASIC LIMITS OF FUNCTIONS AND MAPS

3.3.1. Polynomial-type

Property 1. $\lim_{x\to x_0} a_n x^n + a_{n-1}x^{n-1}+, \cdots, +a_0 = a_n x_0^n + a_{n-1}x_0^{n-1}+, \cdots, +a_0$

Property 2. $\lim_{x\to x_0} \frac{a_n x^n + a_{n-1}x^{n-1}+,\cdots,+a_0}{b_n x^n + b_{n-1}x^{n-1}+,\cdots,+b_0} = \frac{a_n x_0^n + a_{n-1}x_0^{n-1}+,\cdots,+a_0}{b_n x_0^n + b_{n-1}x_0^{n-1}+,\cdots,+b_0},$

$$\forall x \in \mathbb{R}, \text{such that } b_n x^n + b_{n-1}x^{n-1}+, \cdots, +b_0 \neq 0.$$

3.3.2. Trigonometric-type

Property 1. $\lim_{x\to x_0}\sin x = \sin x_0, \forall\ x \in \mathbb{R}.$
Property 2. $\lim_{x\to x_0}\cos x = \cos x_0, \forall\ x \in \mathbb{R}.$
Property 3. $\lim_{x\to x_0}\tan x = \tan x_0, \forall\ x \neq (2z+1)\frac{\pi}{2},\ z \in \mathbb{Z}.$
Property 4. $\lim_{x\to x_0}\cot x = \cot x_0, \forall\ x \neq z\pi,\ z \in \mathbb{Z}.$
Property 5. $\lim_{x\to x_0}\sec x = \sec x_0, \forall\ x \neq (2z+1)\frac{\pi}{2},\ z \in \mathbb{Z}.$
Property 6. $\lim_{x\to x_0}\csc x = \csc x_0, \forall\ x \neq z\pi,\ z \in \mathbb{Z}.$

3.3.3. Exponential-type

Property 1. $\lim_{x\to\infty}e^x = \infty, \forall\ x \in \mathbb{R}.$
Property 2. $\lim_{x\to-\infty}e^x = 0, \forall\ x \in \mathbb{R}.$

3.3.4. Logarithmic-type

Property 1. $\lim_{x\to\infty}\ln x = \infty, \forall\ x \in \mathbb{R}^+.$
Property 2. $\lim_{x\to 0^+}\ln x = -\infty, \forall\ x \in \mathbb{R}^+.$

3.3.5. Important Quote on Limit of a Map

Determining the limit of a map T **is not** resolved by evaluating the map at point t_0. Note that the $\lim_{t\to 2}t^2 = \lim_{t\to 2^-}t^2 = 4$, but $T(t_0) = T(2)$ is not defined. In this case it is not possible to evaluate it.

Remark 3.14. The determination of the limit of a function must always be carried out looking for the tendency of the boundary near the point x_0, *i.e.*, there is no general procedure to obtain it, each case is different.

3.4. CASE STUDY: WHAT IS THE MAXIMUM AREA OF RECTANGLE?

Case 3.1. What will be the maximum area of rectangle that can be obtained with 50cm of wire?

The perimeter of a rectangle of sides x and y, is given by $2x + 2y = 50 = A$.

Solving for $y = 25 - x$, the two sides of the rectangle are x and $25 - x$

Now let's get the function for the area $A(x) = xy = x(25 - x) = 25x - x^2$.

In the graph (Fig. **3.6**) we can see that function $A(x)$ has a maximum value of 12.5. (Rmk. 3.15).

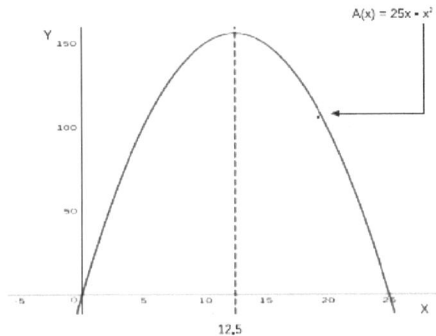

Fig. (3.6). Graph of $25x - x^2$.

Remark 3.15. Note that function A has increasing values in $(-\infty, 12.5)$ and decreasing values in $(12.5, \infty)$.

To get the maximum area we have

$$\lim_{x \to 12.5^+} 25x - x^2 = \lim_{x \to 12.5^-} 25x - x^2 = 156.25.$$

then,

$$\lim_{x \to 12.5} 25x - x^2 = 156.25$$

Hence, the maximum area of rectangle you can get with 50cm of wire is 156.25.

3.5. CASE STUDY: POPULATION GROWTH ESTIMATE

Case 3.2. Suppose that the population growth of a city is expressed with the function f:

$$f(t) = \frac{200 + 15t}{32 + t}$$

.

The year gap from 2010 is t and f expresses **the years**. (i) How many people were there in 2010? (ii) How many people were there in 2016? (iii) Will the population growth be stabilised in the long run? For all cases consider that we are in the year 2019.

(i) There are zero years since 2010, therefore, $f(0) = \frac{200}{32} = 6{,}250$, then in the year 2010 there were 6250 people.

(ii) There are 6 years since 2010, therefore, $f(6) = 7.631578947$, then in the year 2016 there were 7631 people.

(iii) The stabilization will be determined using $\lim_{x \to \infty} \frac{200+15t}{32+t} = 15$ (Rmk. 5).

Therefore, the stabilization will be reached when the neighbourhood gets 15000 people.

Remark 3.16. Dividing by the highest power of the denominator.

$$\lim_{x \to \infty} \frac{200+15t}{32+t} = \frac{\lim_{x \to \infty} \frac{200}{x}+15}{\lim_{x \to \infty} \frac{32}{x}+1} = 15 \qquad (3.3)$$

3.6. CASE STUDY: DRUG ABSORPTION

The amount of drug that stays in the bloodstream after t hours of being introduced is given by the function f. (i) What is the amount of drug that stays in the body? (ii) According to the graph of f (Fig. **3.7**), discuss what is the behaviour of the drug in the body.

$$f(t) = \frac{10t}{t^2 + 1}$$

(i) The amount of drug that stays in the body is determined by $\lim_{x \to \infty} \frac{10t}{t^2+1} = 0$ (Rmk. 3.17).

Therefore, the amount of drug that stays in the body is null.

Remark 3.17. Dividing by the highest power of the denominator.

$$\lim_{t \to \infty} \frac{10t}{t^2+1} = \frac{\lim_{t \to \infty} \frac{1}{x}}{\lim_{t \to \infty} \frac{1}{x^2}+1} = \frac{0}{1} = 0 \tag{3.4}$$

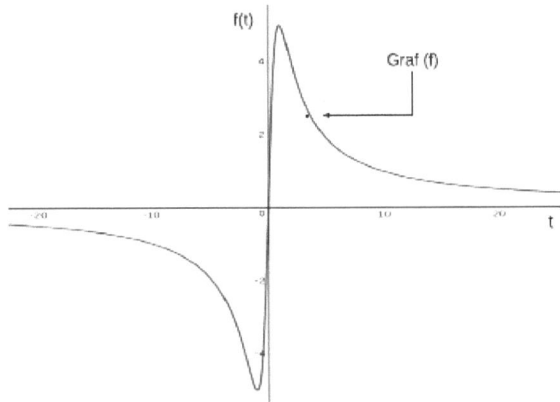

Fig. (3.7). Graph of $\frac{10t}{t^2+1}$.

(ii) The drug rapidly accumulates in the body until it reaches the maximum within the first hour, then it is absorbed and it disappears.

3.7. EXERCISES

Exercise 3.1. Demonstrate using (ε/δ) $\lim_{x\to 2} x^2 + x - 2 = 4$.

Exercise 3.2. Demonstrate using (ε/δ) $\lim_{x\to 2} \frac{1}{x} = \frac{1}{2}$.

Exercise 3.3. Demonstrate using (ε/δ) $\lim_{x\to 0} \frac{1}{x}$ does not exist.

Exercise 3.4. How do we find the limit of a composite function? Be the functions $f: \mathbb{R} \to \mathbb{R}, x^3$, and $g: \mathbb{R}^+ \to \mathbb{R}, \sqrt[4]{x}$. (i) Graphically show the composition of both functions. (ii) Determine $\lim_{x\to 3} f \circ g$.

Exercise 3.5. Find the limit of function $f(x) = [x]$.

Exercise 3.6. Determine the $\lim_{x\to 0} \frac{\sin x}{x}$ [42]. (i) Geometrically represent the relationship between $\sin x$ and x. (ii) Determine the $\lim_{x\to 0} \frac{\sin x}{x}$.

Exercise 3.7. Find the limit of function $f(x) = \frac{\sqrt[3]{27x^3 - 2x + 5}}{x + 1}$ [43], when $x \to \infty$. Graphically show the behaviour of the function.

Exercise 3.8. Find the limit of function $f(x) = \frac{\sqrt[3]{x^2 + 2x - 3}}{|x-1|}$, when $x \to 1^-$ and $x \to 1^+$ [44]. Graphically show the behaviour of the function. Does the limit exist in 1?

Exercise 3.9. Find the limit of function $f(x) = \frac{x-1}{2x^2 + 3}$ [44], when $x \to \infty$. Graphically show the behaviour of the function.

Exercise 3.10. Find the limit of function $f(x) = x\sin\frac{1}{x}$ [44], when $x \to \infty$. Graphically show the behaviour of the function.

CHAPTER 4

Continuity

Abstarct: This chapter introduces the continuity of functions and maps with the limit operator, it also reviews the concept of continuity geometrically and analytically.

Keywords: Bolzano's theorems, Bolzano-Weierstrass' theorem, Continuity of a function, Continuity of a function using one-sided limits, Continuity of a function using two-sided limits, Continuity of a map, continuity of a map using one-sided limits, Continuity of a map using two-sided limits, Functions, Intermediate value theorem, Jump discontinuity of a function, Jump discontinuity of a map, Maps, properties of continuity of a function, Properties of continuity of a map, Removable discontinuity of a function, Removable discontinuity of a map, Weierstrass' theorem over functions, Weierstrass' theorem over maps.

4.1. FUNCTIONS

4.1.1. Continuity of a Function

Definition 4.1. A function f is **continuous at point** $x_0 \in D_f$ if the limit of the function f, as x approaches x_0, exists and is equal to $f(x_0)$

$$\lim_{x \to x_0} f(x) = f(x_0).$$

Remark 4.1. If all points in D_f meet the (Def. 2.1), then function f is continuous.

Example 4.1. Are the following statements true or false? [45] (i) If f is a function such that $\lim f(x)$ as $x \to x_0$ does not exist, then f is not continuous. (ii) If function f is not defined at $x = x_0$, then it is not continuous at $x = x_0$.

Solution 4.1. (i) True. For a function to be continuous at $x = x_0$, $\lim f(x)$ as x is approached, must be equal to $f(x_0)$, the limit must exist and $f(x)$ must be defined at $x = x_0$. (ii) True. See the definition of continuity (Def. 4.7).

Example 4.2. Determine where the function $f(x) = \tan 2x$ is discontinuous [46].

Carlos Polanco
All rights reserved-© 2020 Bentham Science Publishers

Solution 4.2. $f(x) = \tan 2x = \frac{\sin 2x}{\cos 2x}$. Therefore, the denominator will be zero and the function will be discontinuous at $x = \frac{\pi}{4} + n\pi, n \in \mathbb{Z}$.

Example 4.3. The graph of f is given below (Fig. **4.1**). Based on this graph, determine where is the function discontinuous [46].

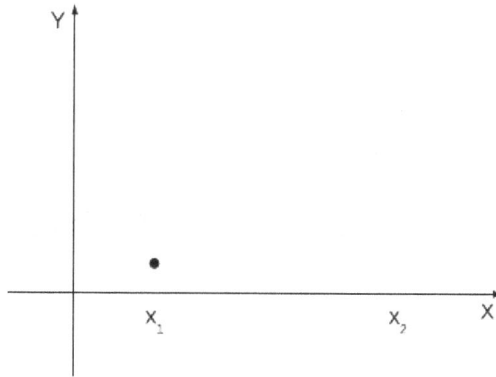

Fig. (4.1). Graphic description of the function f.

Solution 4.3. In x_1, the function is discontinuous because the point $f(x_1)$ is not in the graph of the function. In x_2, the function is discontinuous because the point $f(x_2)$ does not exists.

4.1.2. Properties of Continuity of a Function

Let the functions $f: \mathbb{R} \to \mathbb{R}$ and $g: \mathbb{R} \to \mathbb{R}$ be both continuous at point x_0, If $\lim_{x \to x_0} f(x) = L$, $\lim_{x \to x_0} g(x) = M$ and $\alpha \in \mathbb{R}$, then:

Property 1. $\alpha \lim_{x \to x_0} f(x) = \lim_{x \to x_0} \alpha f(x) = \alpha f(x_0)$ (constant).
Property 2. $\lim_{x \to x_0} f(x) + g(x) = \lim_{x \to x_0} f(x) + \lim_{x \to x_0} g(x) = f(x_0) + g(x_0)$ (addition).
Property 3. $\lim_{x \to x_0} f(x) - g(x) = \lim_{x \to x_0} f(x) - \lim_{x \to x_0} g(x) = f(x_0) - g(x_0)$ (substraction).
Property 4. $\lim_{x \to x_0} f(x) \times g(x) = \lim_{x \to x_0} f(x) \times \lim_{x \to x_0} g(x) = f(x_0) \times g(x_0)$ (multiplication).
Property 5. If $M \neq 0$, $\lim_{x \to x_0} \frac{f(x)}{g(x)} = \frac{\lim_{x \to x_0} f(x)}{\lim_{x \to x_0} g(x)} = \frac{f(x_0)}{g(x_0)}$ (division).

Exmaple 4.4. (i) Determine if the following function is continuous or discontinuous at x = −6. (ii) What about at $x = 1$? (Example taken from [46]) (i)

$$g(x) = \begin{cases} 1 - 3x, & x < -6 \\ 7, & x = -6 \\ x^3, & -6 < x < 1 \\ 1, & x = 1 \\ 2 - x, & x > 1 \end{cases}$$

Solution 4.4. (i) $\lim_{x \to -6^-} g(x) = \lim_{x \to -6^-} (1 - 3x) = \lim_{x \to -6^-} 1 - 3 \lim_{x \to -6^-} x = 1 - 3(-6) = 19$ $\lim_{x \to -6^+} g(x) = \lim_{x \to -6^+} x^3 = (-6)^3 = -216$. The limit does not exist, therefore, this function is not continuous at $x = -6$. (ii) $\lim_{x \to 1^-} g(x) = \lim_{x \to 1^-} x^3 = 1^3 = 1$ $\lim_{x \to 1^+} g(x) = \lim_{x \to 1^+} (2 - x) = \lim_{x \to 1^+} 2 - \lim_{x \to 1^+} x = 2 - 1 = 1$, then $\lim_{x \to 1} g(x) = g(1)$. So the function is continuous at $x = 1$.

4.1.3. Continuity of a Function using One-sided limits

Definition 4.2. Let f be a function defined in the interval $(-\infty, x_0)$, except at point x_0 of that interval, the limit of the function f on x_0 exists, and it is equal to the value L if:

$$\lim_{x \to x_0^-} f(x) = f(x_0)$$

$L \in \mathbb{R}$ is a **left-hand limit of f at x_0**.

Definition 4.3. Let f be a function defined by the interval (x_0, ∞), except at point x_0 of that interval, the limit of the function f on x_0 exists and it is equal to the value L

$$\lim_{x \to x_0^+} f(x) = f(x_0)$$

$L \in \mathbb{R}$ is a **right-hand limit of f at x_0**.

Example 4.5. Determine if the function $f: [1, \infty] \subset \mathbb{R} \to \mathbb{R}, \frac{1}{x}$ is continuous at point $x = 1$.

Solution 4.5. $f(1) = 1$ and $\lim_{x \to 1^+} \frac{1}{x} = 1$, therefore, the function $f(x)$ is continuous at point $x = 1$.

4.1.4. Continuity of a Function using Two-sided Limits

Definition 4.4. From (Def 4.2, 4.3), in a function defined in an interval, except at point x_0 of that interval, the limit of the function f on x_0 exists and it is equal to the value $L \in \mathbb{R}$ if:
$$\lim_{x \to x_0^-} f(x) = f(x_0) = \lim_{x \to x_0^+} f(x)$$
then,

$$\lim_{x \to x_0} f(x) = L$$

$L \in \mathbb{R}$ is a **two-sided limit of f at x_0**.

Example 4.6. Find the point(s) of discontinuity for the function $f(x) = x^2 + 1 + |2x - 1|$ [47].

Solution 4.6. $|2x - 1| = 0 \to x = \frac{1}{2}$, then $f(x) = x^2 + 1 - (2x - 1)$ if $x < \frac{1}{2}$, or $f(x) = x^2 + 1 + (2x - 1)$ if $x \geq \frac{1}{2}$. Since $f(\frac{1}{2}) = \frac{5}{4}$, $\lim_{x \to \frac{1}{2}^-} x^2 + 1 - (2x - 1) = \frac{5}{4}$ and $\lim_{x \to \frac{1}{2}^+} x^2 + 1 - (2x - 1) = \frac{5}{4}$ there are no discontinuities, then the function $f(x)$ is continuous in $\mathbb{R}$.

4.1.5. Removable and Jump Discontinuities of a Function

Definition 4.5. If $\lim_{x \to x_0} f(x) = L$ but $f(x_0)$ is **not** defined or is **different** from L, then the **removable** discontinuity at $x = x_0$ can be removed by defining $f(x_0) = L$.

Example 4.7. Let $f(x) = \frac{x-5}{\sqrt{x-4}-1}$, when $x \neq 5$, f is continuous for all $x > 4$. What is the value of $f(x)$ at point $x = 5$? [48].

Solution 4.7. $\frac{x-5}{\sqrt{x-4}-1} = \frac{(x-5)(\sqrt{x-4}+1)}{(\sqrt{x-4}-1)(\sqrt{x-4}+1)} = \frac{(x-5)(\sqrt{x-4}+1)}{x-5} = \sqrt{x-4} + 1$, then $g(x) = \sqrt{x-4} + 1$. $\lim_{x \to 5^+} \sqrt{x-4} + 1 = 2 = g(5) = 2$.

Definition 4.6. If $\lim_{x \to x_0^+} f(x) = L_1 < \infty$ and $\lim_{x \to x_0^-} f(x) = L_2 < \infty$, but $L_1 \neq L_2$, then the **jump** discontinuity at $x = x_0$ **cannot** be removed.

Example 4.8. From Fig. (**4.2**). (i) What is the $x - value$ at which g has an infinite discontinuity ?. (ii) What is the $x - value$ at which g has a removable discontinuity. (iii) What is the $x - value$ at which g has a jump discontinuity.

Solution 4.8. (i) The value is $x = 3$. (ii) The value is $x = 5$. (iii) The value is $x = -2$.

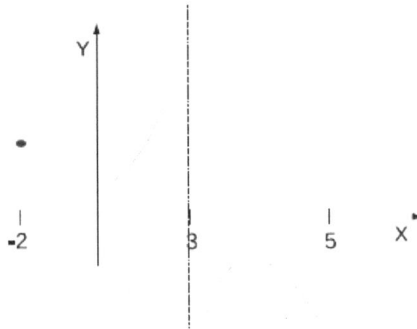

Fig. (4.2). Graphic description of the function g.

4.2. MAPS

4.2.1. Continuity of a Map

Definiution 4.7. A map $T: D \subset \mathbb{R} \to M \subset \mathbb{R}$ is **continuous at point** $t_0 \in D$ if the limit of the map T, as t approaches t_0, exists and is equal to $T(t_0) \in M$

$$\lim_{t \to t_0} T(t) = T(t_0) \in M.$$

Example 4.9. Let the map $T: [0,1] \subset \mathbb{R} \to \mathbb{R}, 2t$. Determine if the map T at point $t = \frac{1}{2}$ is continuous.

Solution 4.9. $\lim_{t \to \frac{1}{2}} T(t) = 2\frac{1}{2} = 1$, on the other hand $T(\frac{1}{2}) = 1$, then the map T is continuous at $t = \frac{1}{2}$.

4.2.2. Properties of Continuity of a Map

Let the map $T_1: \mathbb{R} \to \mathbb{R}$ and $T_2: \mathbb{R} \to \mathbb{R}$ be both continuous at point t_0, if $\lim_{t \to t_0} T_1(t) = L$, $\lim_{t \to t_0} T_2(t) = M$ and $\alpha \in \mathbb{R}$ then:

Property 1. $\alpha \lim_{t \to t_0} T_1(x) = \lim_{t \to t_0} \alpha T_1(x) = \alpha T_1(t_0)$ (constant).

Property 2. $\lim_{t \to t_0} T_1(t) + T_2(t) = \lim_{t \to t_0} T_1(x) + \lim_{t \to t_0} T_2(t) = T_1(t_0) + T_2(t_0)$ (addition).

Property 3. $\lim_{t \to t_0} T_1(t) - T_2(t) = \lim_{t \to t_0} T_1(t) - \lim_{t \to t_0} T_2(t) = T_1(t_0) - T_2(t_0)$ (substraction).

Property 4. $\lim_{t \to t_0} T_1(t) \times T_2(t) = \lim_{t \to t_0} T_1(t) \times \lim_{t \to t_0} T(t) = T_1(t_0) \times T_2(t_0)$ (multiplication).

Property 5. If $M \neq 0$, $\lim_{t \to t_0} \frac{T_1(t)}{T_2(t)} = \frac{\lim_{t \to t_0} T_1(t)}{\lim_{t \to t_0} T_2(t)} = \frac{T_1(t_0)}{T_2(t_0)}$ (division).

4.2.3. Continuity of a Map Using One-sided Limits

Definition 4.8. Let the map $T: D \subset \mathbb{R} \to M \subset \mathbb{R}$ defined in $D = (-\infty, t_0)$, except at point t_0 of that interval. The limit of the map T on t_0 exists and is equal to the value L if:

$$\lim_{t \to t_0^-} T(t) = T(t_0)$$

$L \in (-\infty, x_0) \subset \mathbb{R}$ is a **left-hand limit of T at t_0**.

Example 4.10. Determine if the map $T: [1, \infty] \subset \mathbb{R} \to \mathbb{R}^2, \frac{1}{x}$ is continuous at point $t = 1$.

Solution 4.10. $T(1) = 1$ and $\lim_{t \to 1^+} \frac{1}{t} = 1$, therefore, the map $T(t)$ is continuous at point $t = 1$.

4.2.4. Continuity of a Map using Two-sided Limits

Definition 4.9. Let the map $T: D \subset \mathbb{R} \to M \subset \mathbb{R}$, defined $D = \mathbb{R}$, except at point t_0 of that interval. The limit of the map T on t_0 exists and is equal to the value L if:

$$\lim_{x \to x_0^-} T(t) = T(t_0) = \lim_{x \to x_0^+} T(t)$$

$L \in D \subset \mathbb{R}$ is a **two-sided limit of** T **at** t_0.

Example 4.11. Find the point(s) of discontinuity for the map $T(t) = t^2 + 1 + |2t - 1|$.

Solution 4.11. Since $|2t - 1| = 0 \rightarrow t = \frac{1}{2}$, then $T(t) = t^2 + 1 - (2t - 1)$ if $t < \frac{1}{2}$ or $T(t) = t^2 + 1 + (2t - 1)$ if $t \geq \frac{1}{2}$. As $T(\frac{1}{2}) = \frac{5}{4}$, $\lim_{t \to \frac{1}{2}^-} t^2 + 1 - (2t - 1) = \frac{5}{4}$ and $\lim_{t \to \frac{1}{2}^+} t^2 + 1 - (2t - 1) = \frac{5}{4}$ there is no discontinuities, therefore, the map $T(t)$ is continuous in $\mathbb{R}$.

4.2.5. Removable and Jump Discontinuities of a Map

Definition 4.10. If $\lim t \to t_0 T(t) = L$ but $T(t_0)$ is **not** defined or is **different** to L, then the discontinuity at $t = t_0$ can be **removed** by defining $t(t_0) = L$.

Example 4.12. Let the map $T(t) = \frac{t-5}{\sqrt{t-4}-1}$, when $t \neq 5$, T is a continuous map for all $t > 4$. What is the value of $T(t)$ at point $t = 5$?.

Solution 4.12. $\frac{t-5}{\sqrt{t-4}-1} = \frac{(t-5)(\sqrt{t-4}+1)}{(\sqrt{t-4}-1)(\sqrt{t-4}+1)} = \frac{(t-5)(\sqrt{t-4}+1)}{t-5} = \sqrt{t-4} + 1$. Then $T2t) = \sqrt{t-4} + 1$. $\lim t \to 5^+ \sqrt{t-4} + 1 = 2 = T_2(5) = 2$.

Definition 4.11. If $\lim t \to t_0^+ T(t) = L_1 < \infty$ and $\lim_{t \to t_0^-} T(t) = L_2 < \infty$ but $L_1 \neq L_2$, then the **jump** discontinuity at $t = t_0$ **cannot** be removed.

Example 4.13. From Fig. (**4.2**). (i) Select the $t - values$ where the map T has an infinite discontinuity. (ii) Select the $t - values$ where the map T has a removable discontinuity. (iii) Select the $t - values$ where the map T has a jump discontinuity.

Solution 4.13. (i) The value is $t = 3$. (ii) The value is $t = 5$. (iii) The value is $t = -2$.

4.3. WEIERSTRASS'S THEOREM

Definition 4.12. If a function f is continuous on the closed interval $[a, b]$, then f must attain a maximum and a minimum each at least once. That is, there exist values c and d in $[a, b]$ such that [50]

$$f(c) \leq f(x) \leq f(d), \qquad \forall x \in [a, b]$$

Proof. The set $y \in R: y = f(x)$ forsome $x \in [a, b]$ is a bounded set, hence its **least upper bound** exists by the least upper bound property of the real numbers. Let $M = sup(f(x))$ on $[a, b]$, if there is no point x in $[a, b]$ so that $f(x) = M$, then $f(x) < M$ in $[a, b]$. Therefore, $1/(M - f(x))$ is continuous on $[a, b]$.

However, to every positive value ε, there is always some $x \in [a, b]$ such that $M - f(x) < \varepsilon$, because M is the **least upper bound**. Hence, $1/(M - f(x)) > 1/\varepsilon$, which means that $1/(M - f(x))$ is not bounded. Since every continuous function on $[a, b]$ is bounded, this contradicts the conclusion that $1/(M - f(x))$ was continuous on $[a, b]$. Therefore, there must be a point $x \in [a, b]$ such that $f(x) = M$. (Proof taken from [50]).

Definition 4.13. If a map T is continuous on the closed interval $[a, b]$, then T must attain a maximum and a minimum each at least once. That is, there exist values t_1 and t_2 in $[a, b]$ such that

$$T(t_1) \leq T(t) \leq T(t_2), \qquad \forall t \in [a, b].$$

Proof. The set $y \in R: y = T(t)$ forsome $t \in [a, b]$ is a bounded set, hence its **least upper bound** exists by the least upper bound property of the real numbers. Let $M = sup(T(t))$ on $[a, b]$. If there is no point t on $[a, b]$ so that $T(t) = M$, then $T(t) < M$ on $[a, b]$. Therefore, $1/(M - T(t))$ is continuous on $[a, b]$.

However, to every positive value ε, there is always some $t \in [a, b]$ such that $M - T(t) < \varepsilon$, because M is the **least upper bound**. Hence $1/(M - T(t)) > 1/\varepsilon$, which means that $1/(M - T(t))$ is not bounded. Since every continuous map on a $[a, b]$ is bounded, this is contradictory because $1/(M - T(t))$ was continuous on $[a, b]$. So there is a point $t \in [a, b]$ such that $T(t) = M$.

4.4. BOLZANO'S THEOREM

Definition 4.14. Suppose that f is a continuous function on a closed interval $[a, b]$ and takes the values of the opposite sign at the extremes, and that there is at least one c on (a, b) such that $f(c) = 0$ (Fig. **4.4**).

Proof. Let S be the set of values x within the closed interval from a to b where $f(x) < 0$. Since S is non-empty (it contains a) and S is bounded (it is a subset of $[a, b]$, the least upper bound axiom asserts the existence of a least upper bound, say c, for S (Proof taken from [51]).

Example 4.14. Verify that the equation $x^3 + x - 1 = 0$ has at least one real solution on the interval $[0,1]$ (Example taken from [52]).

Solution 4.14. First consider the function $f(x) = x^3 + x - 1$, which is continuous on $[0,1]$ because it is polynomial. Then, study the signs at the extremes of the interval: $f(0) = -1 < 0, f(1) = 1 > 0.$

As the signs are different Bolzano's theorem can be applied, which determines that there is a c on (0.1) such that $f(c) = 0$. This process demonstrates that there is a solution in this interval.

Definition 4.15. Suppose that T is a continuous map on a closed interval $[a, b]$ that takes the values of the opposite sign at the extremes, and there is at least one t_0 in (a, b) such that $T(t_0) = 0$.

Proof. Let S be the set of values t within the closed interval from a to b where $T(t) < 0$. Since S is non-empty (it contains a) and S is bounded (it is a subset of $[a, b]$), the least upper bound axiom asserts the existence of a least upper bound, say t_0, for S (Proof taken and adapted from [51]).

Example 4.15. Verify that the equation $x^3 + x - 1 = 0$ has at least one real solution in the interval $[0,1]$ (Example taken from [42]).

Solution 4.15. First consider the map $T(t) = t^3 + t - 1$, which is continuous on $[0,1]$ because it is polynomial. Then, study the signs at the extremes of the interval: $T(0) = -1 < 0$, $T(1) = 1 > 0$. As the signs are different, Bolzano's Theorem over Maps can be applied, which determines that there is a t_0 in $[0,1]$ such that $T(t_0) = 0$. This process demonstrates that there is a solution in this interval.

4.5. BOLZANO-WEIERSTRASS'S THEOREM

Theorem 4.1. If f is a continuous function on a closed interval $[a, b]$ and L is any value between $f(a)$ and $f(b)$, then there is at least one value c in $[a, b]$ such that $f(c) = L$.

Proof. If f is continuous on $[a, b]$ then there exists a value $L \in [f(a), f(b)]$ such that $lim_{x \to c} f(x) = L = f(b)$.

Example 4.16. Let f be a continuous function such that $f(-2) = 3$, and $f(1) = 5$. Applying the Intermediate Value Theorem, what can you say about the function? [53].

Solution 4.16. $f(c) = 4$ for at least one value c between -2 and 1.

Theorem 4.2. If T is a continuous map on a closed interval $[a, b]$ and L is any value between $T(a)$ and $T(b)$, then there is at least one value t_0 in $[a, b]$ such that $T(t_0) = L$.

Proof. If T is a continuous map on $[a, b]$, then there exists a value $L \in [T(a), T(b)]$ such that $lim_{t \to t_0} T(t) = L = T(b)$.

Example 4.17. Let T be a continuous map such that $T(-2) = 3$ and $T(1) = 5$. Applying the Bolzano-Weierstrass's Theorem, what can you say about the map? [53].

Solution 4.17. $T(t_0) = 4$ for at least one value t_0 between -2 and 1.

4.6. CASE STUDY: ROOTS OF A POLYNOMIAL FUNCTION

Case 4.1. Since function $x^3 - 15x + 1$ is continuous and at the extremes of the interval $[-4, 4]$ it changes sign: $(-4)^3 - 15(-4) + 1 = -64 + 60 + 1 = -3$ and $(4)^3 - 15(4) + 1 = 64 - 60 + 1 = 5$, this function must have at least a root in the open interval $(4, 4)$ (Theo. 4). However, the theorem does not identify these roots, therefore, we use the bisection method to find the solution.

The method to find a zero in the function f, *i.e.*, $f(c) = 0$ in the interval $[a, b]$ with margin of error e, consists in taking the midpoint of the interval $c = (a + b)/2$ and consider the following rules:

Rule 1. If $|f(c)| < e$, we have found an approximation to the point that cancels f with an acceptable tolerance of e.
Rule 2. If $f(c)$ has different sign than $f(a)$, repeat the procedure in the interval $[a, c]$.
Rule 3. If it does not have different sign, repeat the procedure in the interval $[c, b]$.

We apply this method to the function $x^3 - 15x + 1$, with a tolerance of $e = 1.1$ and calculate $c_1 = (a + b)/2 = (-4 + 4))/2 = 0$. $|f(c_1)| = |f(0)| = 1 > e = 1.1$ the sign is different from $f(-4)$, therefore, our new interval is $[a, c_1] = [-4, 0]$.

We determine $c_2 = (-4 + 0)/2 = (-4))/2 = -2$ for $f(c_2) = f(-2) = -8 + 30 + 1 = 23$, so we repeat the procedure for the interval $[a, c_2] = [-4, -2]$.

Now $c_3 = (-4 - 2)/2 = (-6))/2 = -3$, since $f(c_3) = f(-3) = -27 + 45 + 1 = 19$, we repeat the procedure for the interval $[a, c_3] = [-4, -3]$.

Now $c_4 = (-4 - 3)/2 = (-7))/2 = -3.5$, since $f(c_4) = f(-3.5) = 10.6$, we repeat the same procedure for the interval $[a, c_4] = [-4, -3.5]$.

Now $c_5 = (-4 - 3.5)/2 = (-7.5))/2 = -3.75$, since $f(c_4) = f(-3.75) = 6.59$, we repeat the procedure in the interval $[a, c_5] = [-4, -3.75]$.

Now $c_6 = (-4 - 3.75)/2 = (-7.75))/2 = -3.87$, since $f(c_6) = f(-3.87) = 1.08$.

Note that the last value is inferior to the tolerance e, therefore, the root for this function is -3.87.

4.7. CASE STUDY: VERTICAL AND HORIZONTAL ASYMPTOTES

Case 4.2.

Definition 4.16. Horisontal asymptotes of a function are horizontal lines of the form $y = a$. A function can have at the most two horizontal asymptotes, one at the left (when $x \to -\infty$) and one at the right (when $x \to \infty$). They are calculated in this way:
If $\lim_{x \to -\infty} f(x) = a$, then $y = a$ is a horizontal asymptote for $f(x)$ (at the left).

If $\lim_{x \to +\infty} f(x) = b$, then $y = b$ is a horizontal asymptote of $f(x)$ (at the right).

Definition 4.17. Vertical asymptotes of a function are vertical lines of the form $x = k$. There are no restrictions as to the number of vertical asymptotes that a function can have. There are functions that do not have vertical asymptotes, functions that have only one, two, and even an infinite number of asymptotes. They are calculated in the following way:

If $\lim_{x \to k^-} f(x) = \pm\infty$, then $x = k$ is a vertical asymptote of $f(x)$, at the left if the limit is $-\infty$, or at the right if the limit is $+\infty$).

If $\lim_{x \to k^+} f(x) = \pm\infty$, then $x = k$ is a vertical asymptote of $f(x)$, at the left if the limit is $-\infty$, or at the right if the limit is $+\infty$).

A function can also cross over an asymptote. An example would be the function $f(x) = \frac{\sin(x)}{x}$. This function has a horizontal asymptote $y = 0$ on both sides. We can see (Fig. **4.3**)

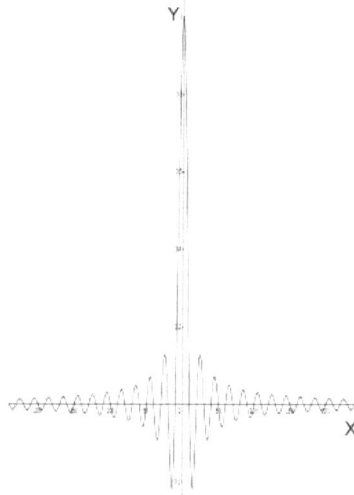

Fig. (4.3). Graphic description of the function $f(x) = \frac{\sin x}{x}$.

A linear function $f(x) = ax + b$ is its own oblique asymptote, such as a constant function $f(x) = k$ is its own horizontal asymptote.

4.8. CASE STUDY: POLLUTION COST

Case 4.3. The cost C, given in millions of euros, of removing an x percentage of the contaminants emitted from a region, can be modelled by

$$C = \frac{3x}{100 - x}$$

(i) What is the implied domain of C? Explain your reasoning. (ii) Use a graphing utility to graph the cost function. Is the function continuous on its domain? Explain your reasoning. (iii) Find the cost of removing 85% of the pollutants from the smokestack. (iv) What is the corresponding map to the graph of the function C?

(i) If x is the percentage of pollutants, then the minimum is 0% and its maximum is 99.99%. Its domain is $[0,99.99]$.

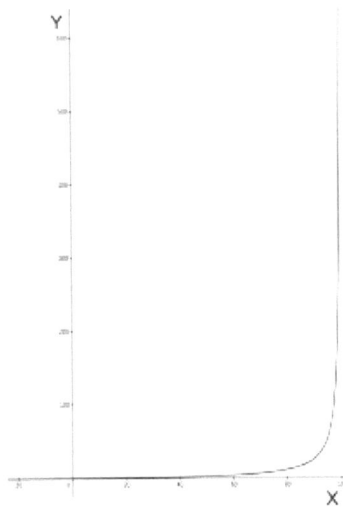

Fig. (4.4). Graph of the function $C(x) = \frac{3x}{100-x}$.

(ii) The graph corresponds to the domain of the function (Fig. **4.4**).

(iii) $C(x) = \frac{3x}{100-x}$ if $x = 85$ then $C(85) = \frac{(3)(85)}{100-85} = 17$. In this case it means 17 million euros.

(iv) $T : [0,100] \subset \mathbb{R} \to \mathbb{R}, \ (t, \frac{3t}{100-t})$.

4.9. EXERCISES

Exercise 4.1. Explain in your own words the meaning of Weierstrass's Theorem over Functions (Theorem 3).

Exercise 4.2. What is the substantive difference between a function and a map?

Exercise 4.3. Explain in your own words the meaning of Bolzano's Theorem over Functions (Theorem 4).

Exercise 4.4. Compute the lim $\lim_{x \to \infty} \frac{x+1}{x^4+2x+2}$.

Exercise 4.5. Explain in your own words the meaning of Bolzano-Weierstrass's Theorem over Functions (Theorem 4.5).

Exercise 4.6. Let the function $f : A \subset \mathbb{R} \Rightarrow B \subset \mathbb{R}, \frac{(x-1)|x-3|}{x^3-4x^2+3x} e^{1-x}$. Study the continuity of the function $g(x) = f(x)$ if $x \in A$; or 0 if $x \notin A$ [54].

Exercise 4.7. Explain in terms of ε/δ the continuity of a function f at point x_0.

Exercise 4.8. Study the continuity of the function [55]

$$f(x) = \begin{cases} x\sin\dfrac{1}{x} & for \quad x \neq 0 \\\\ 1 & for \quad x = 0 \end{cases}$$

Exercise 4.9. Since both $f(x) = x^2 + 1$ and $g(x) = \cos x$ are continuous on $\mathbb{R}$. (i) Study the continuity of $f \circ g$ and (ii) $g \circ f$.

Exercise 4.10. Where is valid the continuity of the function $\frac{\log 1 + x^2}{x^4 - 26x^2 + 25}$ [56]?

CHAPTER 5

Differentiation

Abstarct: This chapter introduces the concept of differentiability of **functions** and **maps**, the properties and rules using the limit operator, the Implicit function theorem, the Inverse function theorem, and L'Hospital's rule.

Keywords: Continuity and differentiability of functions and maps, Derivative of a reciprocal theorem, Differentiation of a function, Differentiation of a map, Functions, Type C^1 function, Implicit function theorem, Inverse function theorem, L'Hospital's rule, Maps, Mean value theorem, Notation for differentiation, Properties of differentiation of functions and maps, Rules of differentiation of functions and maps.

5.1. FUNCTIONS

5.1.1. Derivative of a Function

The **derivative** of a function is the limit of the quotient of an increment of a function and the corresponding increment of a variable, as the increment tends to zero (Fig. **5.1**). **Differentiation** is the action of computing a derivative [57].

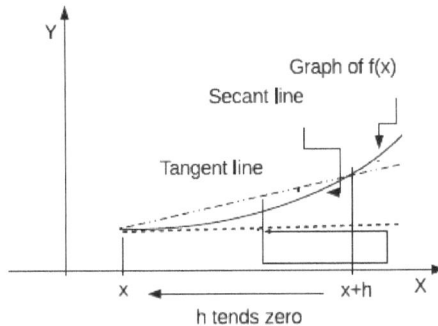

Fig. (5.1). Tangent line as limit of secant lines.

The slope of the line between points $(x, f(x))$ and $(x + h, f(x + h))$ is (Eq. 5.1).

$$\frac{f(x+h)-f(x)}{h} \tag{5.1}$$

Carlos Polanco
All rights reserved-© 2020 Bentham Science Publishers

Definition 5.1. The derivative of a function f is el limit of the slope of the tangent line (Fig. **5.1**). This function (Eq. 5.2) is denoted f' and is called the derivative function or the derivative of f.

$$f'(x) = \lim h \to 0 \frac{f(x+h)-f(x)}{h} \tag{5.2}$$

From (Eq. 5.2) The derivative of function f at point x_0, is (Eq. 5.3)

$$f'(a) = \lim_{h \to 0} \frac{f(a+h)-f(a)}{h} \tag{5.3}$$

Example 5.1. Compute the derivative of $f(x) = x$.

Solution 5.1. $f'(x) = \lim_{h\to 0} \frac{f(x+h)-f(x)}{h} = \lim_{h\to 0} \frac{x+h-x}{h} = \lim_{h\to 0} \frac{h}{h} = 1.$

Example 5.2. Compute the derivative of $f(x) = \frac{x}{x+1}$.

Solution 5.2. $f'(x) = \lim_{h\to 0} \frac{1}{h} f(x+h) - f(x) = \lim_{h\to 0} \frac{1}{h} \frac{h}{(x+h+1)(x+1)} = \frac{1}{(x+1)^2}.$

Example 5.3. Compute the derivative of $f(x) = \cos x$

Solution 5.3. Applying (Rmk. 5.2 and (Ex. 3.10) in (Eq. 5.4), we find that $f'(x) = \sin x$.

$$\lim_{h \to 0} \frac{\cos(x+h)-\cos x}{h} \quad =$$

$$= \lim_{h \to 0} \frac{\cos x \cos h - \sin x \sin h - \cos x}{h}$$

$$= \cos x \lim h \to 0 \frac{\cos h - 1}{h} - \sin x \lim_{h \to 0} \frac{\sin h}{h} \tag{5.4}$$

$$= -\sin x$$

Remark 5.1. $\cos a + b = \cos a \cos b - \sin a \sin b$

Remark 5.2. $\lim_{h \to 0} \frac{\cos h - 1}{h} = 0$

5.1.2. Continuity and Differentiability of Functions

If a function f is differentiable at point x_0 (Def. 5.1) then it is continuous at x_0 (Def. 4.7), but if a function is continuous at that point, it may or may not be differentiable at that particular point.

Example 5.4. Be the function $f(x) = |x|$ over the interval $x \in [-1,1]$. Discuss its continuity.

Solution 5.4. The function f is continuous, but at point 0 its $f'(x)$ is not defined.

Example 5.5. Consider the function $f(x) = (3x - 2)^{\frac{1}{3}}$. Discuss its continuity and differentiability at $x_0 = \frac{2}{3}$.

Solution 5.5. To check the **continuity**, we verify the left hand and right-hand limits and the value of the function at point $x_0 = \frac{2}{3}$.

$$\lim_{x \to \frac{2}{3}^-} (3x - 2)^{\frac{1}{3}} = 0 = \lim_{x \to \frac{2}{3}^+} (3x - 2)^{\frac{1}{3}}$$

Now, to check the **differentiability** at the given point

$$\lim_{h \to 0} \frac{f(x + h) - f(x)}{h} = \lim_{h \to 0} \frac{(3[x + h] - 2)^{\frac{1}{3}} - (3x - 2)^{\frac{1}{3}}}{h} \Big|_{x_0} = \lim_{h \to 0} \sqrt[3]{3} \frac{1}{h^{\frac{2}{3}}} = \infty.$$

Thus, the function $f(x)$ is **not** differentiable at $x_0 = \frac{2}{3}$. See (Fig. **5.2**)

Remark 5.3. It is important to note that the limit in the **continuity** of the function is different than the limit in the **derivative** of the function.

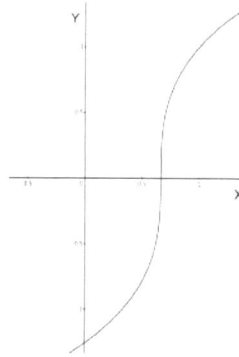

Fig. (5.2). Graph of function $f(x) = (3x - 2)^{\frac{1}{3}}$.

5.1.3. Properties of Differentiation of Functions

Let the functions $f: \mathbb{R} \to \mathbb{R}$ and $g: \mathbb{R} \to \mathbb{R}$ be both differentiable functions at point x_0, then the following properties hold:

Property 1. $f(x) = x^n \Rightarrow f'(x) = nx^{n-1}$ (polynomial)
Property 2. $(\alpha f(x))'_x(x_0) = \alpha \, f'(x_0)$, if $\alpha \in \mathbb{R}$ (constant).
Property 3. $(f + g)'_x(x_0) = f'(x_0) + g'(x_0)$ (addition).
Property 4. $(f - g)'_x(x_0) = f'(x_0) - g'(x_0)$ (substraction).
Property 5. $(f \; g)'_x(x_0) = f'(x_0) \; g(x_0) + f(x_0) \; g'(x_0)$ (multiplication).
Property 6. $\left(\frac{f(x_0)}{g(x_0)}\right)'_x = \frac{f'(x_0)g(x_0)-f(x_0)g'(x_0)}{g^2(x_0)}$ (division).
Property 7. $(f \circ g)'_x(x_0) = f'(y_0) \; g'(x_0)$, where $y_0 = g(x_0)$ (composition).

Remark 5.4. In this work, we will use three different notations for the derivatives of a function f, Leibniz's notation [59] $\frac{dy}{dx}$, Lagrange's notation [60] $f'(x)$, and $(f(x))'$, which is the most commonly used. In case that the derivative is with respect to time t, the notation $\dot{x} = \frac{df}{dt}$ is used. This notation is preferably used in differential equations [61].

A function $f(x)$ can accept higher order derivatives e.g. $f(x) = \sin x$ accepts $f'(x) = \cos x$, $f''(x) = -\sin x$, $f'''(x) = -\cos x$, $f^{iv}(x) = \sin x$. In general, this function accepts n order derivatives. An alternative notation can be $f''(x) = \frac{d^2 f}{dx^2}$, $f'''(x) = \frac{d^3 f}{dx^3}$.

Example 5.6. Be the functions $f(x) = x^3$ and $g(x) = \sqrt{x}$. Compute the properties (Prop. 5.1.3).

Solution 5.6.
(i)　$f'(x) = 3x^2$.　(ii)　$3(f(x))' = 3(3x^2) = 9x^2 = (3f(x))'$.　(iii)　$(f(x) + g(x))' = (x^3 + \sqrt{x})' = 3x^2 + \frac{1}{2\sqrt{x}} = (x^3)' + (\sqrt{x})'$. (iv) $(f(x) - g(x))' = (x^3 - \sqrt{x})' = 3x^2 - \frac{1}{2\sqrt{x}} = (x^3)' - (\sqrt{x})'$.　(v)　$(f(x)g(x))' = (3x^2)(\sqrt{x}) + (x^3)(\frac{1}{2\sqrt{x}}) = 3x + \frac{x^{\frac{5}{2}}}{2}$.　(vi)　$(f(x)/g(x))' = \frac{f'(x)g(x) - f(x)g'(x)}{g^2(x)} = \frac{3x^2\sqrt{x} - x^3 \frac{1}{2\sqrt{x}}}{x} = x^{\frac{3}{2}} - \frac{x^{\frac{3}{2}}}{2} = \frac{x^{\frac{3}{2}}}{2}$. (vii) $(f \circ g)' = ([\sqrt{x}]^3)' = (x^{\frac{3}{2}})' = \frac{3}{2\sqrt{x}}$.

Remark 5.5. In general, the composition is not commutative $(f \circ g)' \neq (g \circ f)'$.

Definition 5.2. A function whose derivative is continuous is said to be of class C^1. Any C^1 function is differentiable [62].

Example 5.7. Provide an example of function $f: \mathbb{R} \to \mathbb{R}$ of class C^1, and another that is not.

Solution 5.7. Function $f(x) = x^2$ is class C^1 because its f' is continuous. Function $f(x) = \sqrt{x}$ is **not** C^1 because $f'(x) = \frac{1}{2\sqrt{x}}$ is not continuous at point 0.

Remark 5.6. If the second derivative of a function f is continuous and exists, then it is of class C^2.

5.1.4. Derivatives of Main Functions

From (Def. 5.1), all the derivatives of the main functions have been calculated, some of them are explained here.

Definition 5.3. (*Polynomial-type*) The derivative of a function $f(x) = a_0 + a_1 x^1 + a_2 x^2 \cdots a_n x^n$, where $n \in \mathbb{N}$, and $a_i \in \mathbb{R}$ is $f'(x) = a_1 x^0 + 2a_2 x^1 \cdots n a_n x^{n-1}$.

Example 5.8. Compute the derivative of the function $f(x) = c$ where $c \in \mathbb{R}$, using the limit definition.

Solution 5.8. $f'(x) = \lim_{h \to 0} \frac{f(x+h)-f(x)}{h} = \lim_{h \to 0} \frac{c-c}{h} = 0.$

Example 5.9. Compute the derivative of the function $f(x) = (2x-3)^2$ at point $x = 2$, using the differentiation properties.

Solution 5.9. By the composition property $f'(x) = 2(2x-3)(2x-3)' = 2(x-3)(2) = 4x-12.$, then $f'(2) = -4.$

Definition 5.4. (*Trigonometric type*) The derivative of the trigonometric functions and their inverse functions are: $f(x) = \sin x \Rightarrow f'(x) = \cos x$, $f(x) = \cos x \Rightarrow f'(x) = -\sin x$, $f(x) = \tan x \Rightarrow f'(x) = \sec^2 x$, $f(x) = \cot x \Rightarrow f'(x) = -\csc^2 x$, $f(x) = \sec x \Rightarrow f'(x) = \sec x \tan x$, $f(x) = \csc x \Rightarrow f'(x) = -\cot x \csc x$, $f(x) = \sin^{-1} x \Rightarrow f'(x) = \frac{1}{\sqrt{1-x^2}}$, $f(x) = \cos^{-1} x \Rightarrow -f'(x) = \frac{1}{\sqrt{1-x^2}}$, $f(x) = \tan^{-1} x \Rightarrow f'(x) = \frac{1}{1+x^2}$, $f(x) = \cot^{-1} x \Rightarrow f'(x) = -\frac{1}{1+x^2}$, $f(x) = \sec^{-1} x \Rightarrow f'(x) = \frac{1}{|x|\sqrt{1-x^2}}$, and $f(x) = \csc^{-1} x \Rightarrow f'(x) = -\frac{1}{|x|\sqrt{1-x^2}}.$

Example 5.10. Compute the derivative of the function $f(x) = \cos 2x^2$, using the differentiation properties.

Solution 5.10. By (Prop. 5.1.3 vii) $f'(x) = -2x \sin 2x^2.$

Example 5.11. Compute the derivative of the function $f(x) = \frac{\sin x}{1+\cos x}$ using the differentiation properties.

Solution 5.11. $f'(x) = \frac{\cos x(1+\cos x) - \sin x(-\sin x)}{(1+\cos x)^2} = \frac{1}{1+\cos x}.$

Definition 5.5. (*Exponential type*) The derivative of a function $f(x) = e^x$ is $f'(x) = e^x$ and the derivative of a function $f(x) = a^x$ is $f'(x) = a^x \ln a.$

Example 5.12. Compute the derivative of the function $f(x) = e^{x^2}$ using the differentiation properties.

Solution 5.12. By the composition property and the definition of the exponential type derivative (Def. 5.5), $f'(x) = 2xe^{x^2}.$

Example 5.13. Show that the function $f(x) = e^{-x}\sin x$ satisfies $\frac{d^2y}{dx^2} + \frac{2dy}{dx} + 2y = 0$.

Solution 5.13. Since $f'(x) = e^{-x}[\cos x - \sin x], f''(x) = e^{-x}[-2\cos x]$, then $e^{-x}[\cos x - \sin x] + 2e^{-x}[-2\cos x] + 2e^{-x}\sin x = 0$.

Example 5.14. Compute the derivative of the function $f(x) = 2^{x^2}$ using the definition of exponential type derivative (Def. 5.5.).

Solution 5.14. $f'(x) = 2^{x^2}\ln 2$.

Example 5.15. Compute the derivative of the function $f(x) = 4^x$ using the definition of exponential type derivative (Def. 5.5).

Solution 5.15. $f'(x) = 4^x\ln 4$.

Definition 5.6. (*Logarithmic type*) The derivative of a function $f(x) = \ln x$ is $f'(x) = \frac{1}{x}$ and if $f((x) = \log_a x = \frac{1}{x\ln x}$, then $\log_a x = \frac{\ln x}{\ln a}$.

Example 5.16. Compute the derivative of the function $f(x) = 4^x - 5\log_9 x$, using the definition of exponential type derivative (Def. 5.6).

Solution 5.16. $f'(x) = 4^x\ln 4 - \frac{5}{x\ln 9}$.

Example 5.17. Compute the derivative of the function $f(x) = 3e^x + 10x^3\ln x$, using the definition of exponential type derivative (Def. 5.6).

Solution 5.17. $f'(x) = 3e^x + 30x^2\ln x + 10x^3\frac{1}{x} = 3e^x + 30x^2\ln x + 10x^2$.

2.2. MAPS

2.2.1. Differentiation of a Map

The **derivative** of a map is the limit of the quotient of an increment of a map and the corresponding increment of a variable t, as the increment tends to zero (Fig. 5.3). **Differentiation** is the action of computing a map [57].

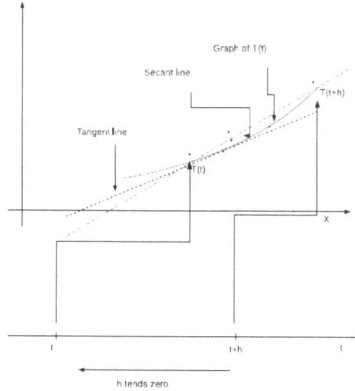

Fig. (5.3). Tangent line as limit of secant lines.

The slope of the line between points $(t, T(t))$ and $(t + h, T(t + h))$ is (Eq. 5.5).

$$\frac{T(t+h)-T(t)}{h} \qquad (5.5)$$

Definition 5.7. The derivative of a map $T: U \subset \mathbb{R} \to M \subset \mathbb{R}^2, (T_1, T_2))$ is the limit of both real valued functions $T_1(t)$ and $T_2(t)$, when $t \to t_0$ (Fig. **5.1**). This map (Eq. 5.6) is denoted T' and is called the derivative map or the derivative of T.

$$T'(t) = (\lim_{h \to 0} \frac{T_1(t+h)-T_1(t)}{h}, \lim_{h \to 0} \frac{T_1(t+h)-T_2(t)}{h}) \qquad (5.6)$$

Example 5.18. Compute the derivative of $T(t) = (t, t)$.

Solution 5.18. $T'(t) = (\lim_{h \to 0} \frac{T(t+h)-T(t)}{h}, \lim_{h \to 0} \frac{T(t+h)-T(t)}{h}) = \lim_{h \to 0}(\frac{h}{h}, \frac{h}{h}) = (1,1)$.

Example 5.19. Compute the derivative of $T(t) = (t, t^2)$.

Solution 5.19. $\lim_{h \to 0}(\frac{T_1(t+h)-T_1(t)}{h}, \frac{T_2(t+h)-T_2(t)}{h}) = \lim_{h \to 0}(\frac{ht}{h}, \frac{(t+h)^2-t^2}{h}) = (\lim_{h \to 0}\frac{h}{h}, \lim_{h \to 0}\frac{h(2t+h)}{h}) = (\lim_{h \to 0}1, \lim_{h \to 0}2t) = (1,2t)$.

5.2.2. Continuity and Differentiability of Maps

If a map T is differentiable at point t_0 (Def. 5.7), then it is continuous at t_0 (Def. 4.7), but if a map is continuous at that point, it may or may not be differentiable at that point.

Example 5.20. Let the map $T(t) = (t, |2t|)$ over the interval $x \in [-1,1]$. Discuss its continuity.

Solution 5.20. The map T is continuous, but at point 0 its $T'(t)$ is not defined.

Example 5.21. Consider the map $T(t) = (t, (3t - 2)^{\frac{1}{3}})$. Discuss its continuity and differentiability at $t_0 = \frac{2}{3}$.

Solution 5.21. To check the **continuity**, we verify the left-hand and right-hand limits of the second component of the map (the first component of the map has already been verified *e.g.* 46), and the value of the map at point $t_0 = \frac{2}{3}$.

$$\lim_{t \to \frac{2}{3}^-}(3t - 2)^{\frac{1}{3}} = 0 = \lim_{x \to \frac{2}{3}^+}(3t - 2)^{\frac{1}{3}}$$

Now, we check the **differentiability** at the given point,

$$T'_2(t) = \lim_{h \to 0}\frac{T(t+h) - T(t)}{h} = \lim_{h \to 0}\frac{(3[t+h] - 2)^{\frac{1}{3}} - (3t - 2)^{\frac{1}{3}}}{h}\Big|_{t_0} = \lim_{h \to 0}\sqrt[3]{3}\frac{1}{h^{\frac{2}{3}}} = \infty.$$

The limit does not exist, therefore, the map $T(t)$ is **not** differentiable at $t_0 = \frac{2}{3}$. See (Fig. **5.2**)

Remark 5.7. It is important to note that the limit in the **continuity** of the map is different than the limit of the **derivative** of the map.

5.2.3. Properties of Differentiation of Maps

Let the maps $T_1 \colon \mathbb{R} \to \mathbb{R}^2$ and $T_2 \colon \mathbb{R} \to \mathbb{R}^2$ be both differentiable at point t_0, then the following properties hold:

Property 1. $T(t) = (t, t^n) \Rightarrow T'(t) = (1, nt^{n-1})$ (polynomial)
Property 2. $(\alpha T(t))'(t_0) = \alpha\, T'(t_0)$, if $\alpha \in \mathbb{R}$ (constant).
Property 3. $(T_1 + T_2)'(t_0) = T_{1'}(t_0) + T_{2'}(t_0)$ (addition).
Property 4. $(T_1 - T_2)'(t_0) = T_{1'}(t_0) - T_{2'}(t_0)$ (substraction).
Property 5. $(T_1\, T_2)'(t_0) = T_{1'}(t_0)\, T_2(t_0) + T_1(t_0)\, T_{2'}(t_0)$ (multiplication).
Property 6. $(\frac{T_1(t_0)}{T_2(t_0)})' = \frac{T_{1'}(t_0)T_2(t_0) - T_1(t_0)T_{2'}(t_0)}{T_2^2(t_0)}$ (division).
Property 7. $(T_1 \circ T_2)'(t_0) = T_{1'}(y_0)\, T_{2'}(t_0)$, where $y_0 = T_2(t_0)$ (composition).

Remark 5.8. In this work, we will use three notation forms for the derivative of a map: Leibniz's notation [59] $T'(t) = \frac{dT}{dt}$, Lagrange's notation ? $T'(t)$, and the notation $(T(t))'_x$, which is the most commonly used.

Example 5.22. Be the maps $T_1(t) = (t^3)$ and $T_2(t) = (\sqrt{x})$. Compute the properties (Prop. 5.2.3).

Solution 5.22.

(i) $T_{1'}(t) = (3t^2)$. (ii) $3(T_1(t))' = (3(3t^2)) = (9t^2) = (3T_1(t))'$. (iii) $(T_1(t) + T_2(t))' = ((t^3 + \sqrt{t})') = (3t^2 + \frac{1}{2\sqrt{t}}) = ((t^3)' + (\sqrt{t})')$. (iv) $(T_1(t) - T_2(t))' = ((t^3 - \sqrt{t})') = (3t^2 - \frac{1}{2\sqrt{t}}) = (t^3)' - (\sqrt{t})'$. (v) $(T_1(x)T_2(x))' = ((3t^2)(\sqrt{t}) + (t^3)(\frac{1}{2\sqrt{t}})) = (3t + \frac{t^{\frac{5}{2}}}{2})$. (vi) $(T_1(t)/T_2(t))' = (\frac{T_{1'}(t)T_2(t) - T_1(t)T_{2'}(t)}{T_2^2(x)}) = (\frac{3t^2\sqrt{t} - t^3\frac{1}{2\sqrt{t}}}{t}) = (t^{\frac{3}{2}} - \frac{t^{\frac{3}{2}}}{2}) = (\frac{t^{\frac{3}{2}}}{2})$. (vii) $(T_1 \circ T_2)' = (([\sqrt{t}]^3)') = ((t^{\frac{3}{2}})') = (\frac{3}{2\sqrt{t}})$.

Remark 5.9. In general, the composition of functions is not commutative $(T_1 \circ T_2)' \neq (T_2 \circ T_1)'$.

Definition 5.8. A map whose derivative is continuous is of class C^1. Any C^1 map is differentiable.

Example 5.23. Provide an example of map $T: \mathbb{R} \to \mathbb{R}$ that is of class C^1 and an example of a map that is not.

Solution 5.23. Map $T(t) = (t^2)$ is class C^1 because its $T'(t)$ is continuous. Map $T(t) = (\sqrt{t})$ is not C^1 because $T'(t) = (\frac{1}{2\sqrt{t}})$ is not continuous at point 0.

Remark 5.10. If the second derivative of the map T is continuous and exists, then it is of class C^2.

5.3. DERIVATIVE TEST AND CRITICAL POINTS

This test and definition of the critical points apply for both functions and maps.

Definition 5.9. If a function f is C^2 (Def. 1.3) at a **critical point** x i.e. $f'(x) = 0$, then:

    1.   If $f''(x) < 0$, then f has a **local maximum** at x.
    2.   If $f''(x) > 0$, then f has a **local minimum** at x.
    3.   If $f''(x) = 0$, then the test is inconclusive.

Example 5.24. Let the function type C^2 $f(x) = x^2$. (i) Determine its critical points. (ii) Are these critical points maximum or minimum?

Solution 5.24. (i) $f'(x) = 2x$, then $2x = 0 \Rightarrow x = 0$, there is only one critical point. (ii) $f''(x) = 2 > 0$, therefore, the function reaches a maximum at point $x = 0$.

Example 5.25. Let the function type C^2 $f(x) = xe^{x^2}$. Determine the critical points.

Solution 5.25. $f'(x) = e^{x^2}(1 + 2x^2)$, then $e^{x^2}(1 + 2x^2) = 0$. In this function, any real value of x will be zero, therefore, there are no critical points.

Example 5.26. Be the map type C^2 $T(t) = (\sin t), t \in [0, 2\pi]$. Determine its critical points.

Solution 5.26. $T'(t) = (\cos t)$, then $\cos t = 0 \Rightarrow t = \frac{\pi}{2}, \frac{3\pi}{2}$, both values are critical points of the map T. Thus $T''(\frac{\pi}{2}) = (-\sin\frac{\pi}{2})$ is less than zero and this implies $\frac{\pi}{2}$ is a local maximum. $T''(\frac{3\pi}{2}) = -\sin\frac{3\pi}{2} > 0$, so $\frac{\pi}{2}$ is a local minimum.

Remark 5.11. When the method is not conclusive, it is necessary to observe how the signs and values of the function or the map change in the interval near the critical point.

5.4. MEAN VALUE THEOREM

This section applies for functions or maps.

Theorem 5.1. Given a continuous function f defined by a closed interval $[a, b]$ and differentiable on the open interval (a, b), where $f(a) = f(b)$, there is a value c such that $a < c < b$ and $f'(c) = \frac{f(b)-f(a)}{b-a}$.

Proof. Let's define an auxiliary function g such that $g(x) = f(x) - f(a) - k(x - a)$, $k \in \mathbb{R}$, then $g'(x) = f'(x) - k$. If $x = a$, $g(a) = 0$, $x = b$ then $g(b) = 0$. Thus $k = \frac{f(b)-f(a)}{b-a}$. Here we have a function g that meets what function f states, therefore, $g'(c) = f'(c)$. So $f'(c) = \frac{f(b)-f(a)}{b-a}$.

Example 5.27. Determine all the values c that satisfy the conclusions of the Mean Value Theorem for: (i) the function $f(x) = x^3 + 2x^2 - x$ on $[-1,2]$ (Example taken from ?), (ii) the map $T: [-1,1] \subset \mathbb{R} \Rightarrow \mathbb{R}$, $(t^2 - t + 3)$.

Solution 5.27. (i) $f'(x) = 3x^2 + 4x$, it implies that $f'(c) = \frac{f(2)-f(-1)}{2-(-1)} \Leftrightarrow 3c^2 + 4c - 1 = \frac{14-2}{3} = 4 \Rightarrow c = \frac{-4+\sqrt{76}}{6}$. (ii) $T'(t) = (2t - 1)$, it implies that $T'(c) = (\frac{T(1)-T(-1)}{1-(-1)}) \Leftrightarrow 2c - 1 = \frac{3-5}{2} = -1 \Rightarrow c = 0$.

5.5. L'HOSPITAL'S RULE

This section applies for functions or maps.

Theorem 5.2. Type $\frac{0}{0}$: ?, ? For $[a, b]$ an interval $x_0 \in [a, b]$ and $f, g: [a, b] \setminus \{x_0\} \subset \mathbb{R} \to \mathbb{R}$, the functions have to meet these conditions: (i) the derivatives of f and g exist in $[a, b] \setminus \{x_0\}$. (ii) $g'(x) \neq 0 \; \forall x \in [a, b] \setminus \{x_0\}$. (iii) $\lim_{x \to c} f(x) = \lim_{x \to x_0} g(x) = 0$, then

$$\lim_{x \to x_0} \frac{f'(x)}{g'(x)} = L \in \mathbb{R} \Rightarrow \lim_{x \to x_0} \frac{f(x)}{g(x)} = L.$$

Proof. If $g'(x_0) = 0$ and $g'(x) \neq 0$, then $x \neq 0$ (Thm. 4) $g(x) \neq 0$; and if $f(x_0) = g(x_0) = 0$ (Thm. 4) $\forall x \in (a, b)$, with $x \neq x_0$ then $\exists \; x_t \in [a, b]$ such that

$$\frac{f(x)}{g(x)} = \frac{f(x) - f(x_t)}{g(x) - g(x_t)} = \frac{f'(x_t)}{g'(x_t)}.$$

Therefore, $\lim_{x \to x_0} \frac{f(x)}{g(x)} = \lim_{x \to x_0} \frac{f'(x_t)}{g'(x_t)} = L$.

Example 5.28. Let the functions $f(x) = x^2 - 9$, $g(x) = x - 3$, and a point $x_0 = 3$. Compute the $\lim_{x \to 0} \frac{f(x)}{g(x)}$.

Solution 5.28. The functions f and g comply with (Thm. 5, i–iii). So $\lim_{x \to a} \frac{f'(x)}{g'(x)} = \lim_{x \to 0} \frac{x^2 - 9}{x - 3} \left[\frac{0}{0}\right] = \lim_{x \to 3} \frac{2x}{1} \left[\frac{0}{0}\right] = \lim_{x \to 3} \frac{6}{1} \left[\frac{0}{0}\right] = 6$.

Example 5.29. Let the maps $T_1(t) = (\sin t)$, $T_2(t) = (t)$, and a point $x_0 = 0$. Compute the $\lim_{t \to 0} \frac{T_1(t)}{T_2(t)}$.

Solution 5.29. The maps T_1 and T_2 comply with (Thm. 5i–iii), so $\lim_{t \to t_0} \frac{T_{1'}(t)}{T_{2'}(t)} = (\lim_{t \to 0} \frac{\sin t}{t} \left[\frac{0}{0}\right]) = (\lim_{t \to 0} \frac{\cos t}{1} \left[\frac{0}{0}\right]) = (\lim_{t \to 0} \frac{1}{1} \left[\frac{0}{0}\right]) = (1)$.

Theorem 5.3. Type $\frac{\infty}{\infty}$: For $[a, b]$ an interval $x_0 \in [a, b]$ and $f, g: [a, b] \backslash \{x_0\} \subset \mathbb{R} \to \mathbb{R}$, the functions have to meet: (i) the derivatives of f and g exist in $[a, b] \backslash \{x_0\}$. (ii) $g'(x) \neq 0 \; \forall x \in [a, b] \backslash \{x_0\}$. (iii) $\lim_{x \to x_0} f(x) = \pm \infty$, and $\lim_{x \to x_0} g(x) = \pm \infty$ then:

$$\lim_{x \to x_0} \frac{f'(x)}{g'(x)} = \lim_{x \to x_0} \frac{f(x)}{g(x)}.$$

Proof. If $g'(x_0) = 0$ and $g'(x) \neq 0$, then $x \neq 0$ (Thm. 4) $g(x) \neq 0$; and if $f(x_0) = g(x_0) = 0$ (Thm. 4) $\forall x \in (a, b)$, with $x \neq x_0$, then $\exists \; x_t \in [a, b]$ such that

$$\frac{f(x)}{g(x)} = \frac{f(x) - f(x_t)}{g(x) - g(x_t)} = \frac{f'(x_t)}{g'(x_t)}.$$

Therefore, $\lim_{x \to x_0} \frac{f(x)}{g(x)} = \lim_{x \to x_0} \frac{f'(x_t)}{g'(x_t)} = L$.

Example 5.30. Let the functions $f(x) = \ln x$, $g(x) = x^2$, and a point $x_0 = \infty$. Compute the $\lim_{x \to \infty} \frac{f(x)}{g(x)}$.

Solution 5.30. The functions f and g comply with (Thm. 5 i–iii), so $\lim_{x \to x_0} \frac{f'(x)}{g'(x)}$

$$= \lim_{x \to \infty} \frac{\ln x}{x^2}_{[\frac{\infty}{\infty}]} = \lim_{x \to \infty} \frac{\frac{1}{x}}{2x}_{[\frac{\infty}{\infty}]} = \lim_{x \to \infty} \frac{1}{2x^2}_{[\frac{\infty}{\infty}]} = 0.$$

Example 5.31. Let the maps $T_1(t) = (t)$, $T_2(t) = (\ln t + 2t)$, and a point $t_0 = \infty$.
Compute $\lim_{t \to \infty} \frac{T_1(t)}{T_2(t)}$.

Solution 5.31. The maps T_1 and T_2 comply with (Thm. 5 i–iii), so $\lim_{t \to t_0} \frac{T_{1\prime}(t)}{T_{2\prime}(t)} =$

$(\lim_{t \to \infty} \frac{t}{\ln t + 2t}_{[\frac{\infty}{\infty}]}) = (\lim_{t \to \infty} \frac{1}{\frac{1}{t}}_{[\frac{\infty}{\infty}]}) = (\lim_{t \to \infty} t_{[\frac{\infty}{\infty}]}) = (\infty).$

Remark 5.12. According to what we have studied so far, this limit does not exist.

5.6. Implicit Differentiation

A function $f(x)$ can be **explicitly** defined, this means that a function f is isolated or clear *e.g.* $f(x) = 1 - x$; or **implicitly** defined in an equation *e.g.* $x + f(x) - 1 = 0$. When it is an implicit function, it is not always easy to see what is its **domain**, its **image**, or even its **derivative**. Sometimes it is not possible to define an equation as an explicit function. This concept applies for functions or maps *i.e.* $t + T(t) - 1 = 0$.

Example 5.32. Let the equation $x^2 + y = 1$. (i) Implicitly define the function $f(x)$. (ii) Explicitly define the function $f(x)$. (iii) Verify (ii).

Solution 5.32. (i) $g(x, y) = x^2 + y - 1 = 0$ or $x^2 + f(x) - 1 = 0$. (ii) $f(x) = 1 - x^2$. (iii) If we substitute (ii) in (i), we get $x^2 + f(x) = 1 \Rightarrow x^2 + y - 1 = 0 \Rightarrow x^2 + (1 - x^2) - 1 = 0$.

Example 5.33. Analyse and discuss the implicit and explicit expressions of the equation $x^2 + y^2 = 1$.

Solution 5.33. The **explicit** expression **does not** correspond to a function $f(x) = \pm\sqrt{1 - x^2}$ (Ex. 1). If each positive (+) or negative (−) options are taken separately, then there is an **explicit** function $x^2 + y^2 - 1 = 0$ for both of them. Thus, for each function f_+ and f_- the domain is $D_f = [-1,1]$ and the image $I_f = [-1,1]$. Let's determine the derivative of both functions: $f'_+ = -\dfrac{x}{\sqrt{1-x^2}}$ and $f'_- = \dfrac{x}{\sqrt{1-x^2}}$. Note that both derivatives are only defined in the open interval $(-1,1)$ and not in the closed interval originally mentioned, therefore, if we want any of the functions have their respective derivatives, we will have to restrict their domain to $D_f = (-1,1)$.

The **derivative** of a function (Sect. 6) is the **composition derivative** (Prop. 5.1.3) of each term (Ex. 5.36).

Example 5.34. Let the equation $x^2 + y^3 - 1 = 0$. (i) Obtain $\dfrac{dy}{dx}$ with the derivative of the composition. (ii) Obtain $\dfrac{dT}{dt}$ with the derivative of the composition.

Solution 5.34.

(i)

$$\frac{d}{dx}(x^2 + y^3 - 1) = 0$$

$$\frac{d(x^2)}{dx} + \frac{d(y^3)}{dx} - \frac{d(1)}{dx} = 0$$

$$2x\frac{dx}{dx} + 3y^2\frac{dy}{dx} - 0 = 0 \qquad\qquad (5.7)$$

$$2x + 3y^2\frac{dy}{dx} = 0$$

$$\frac{dy}{dx} = -\frac{2x}{3y^2}$$

(ii)

$$\frac{d}{dt}(t^2 + T^3 - 1) = 0$$

$$\frac{d(t^2)}{dt} + \frac{d(T^3)}{dt} - \frac{d(1)}{dt} = 0$$

$$2t\frac{dt}{dt} + 3T^2\frac{dT}{dt} - 0 = 0 \qquad\qquad (5.8)$$

$$2t + 3T^2\frac{dT}{dt} = 0$$

$$\frac{dT}{dt} = -\frac{2t}{3T^2}$$

5.7. IMPLICIT FUNCTION THEOREM

With this theorem, we can determine the domain of f where its derivative f' exists. Its application will be reviewed in Taylor's series (Ch. 7). This section applies for functions or maps.

An implicit function (Def. 5.6) can be seen as a two variable function $f(x, y)$, whose derivative with respect to one variable implies that the other variable is held constant. For instance, with the implicit function $xy = 0$, we can get the function $f(x, y) = xy$ and its derivative with respect to x, which implies that y will be constant.

Definition 5.10. Given a function $f : \mathbb{R}^2 \to \mathbb{R}$, its **partial derivatives** with respect to one of those variables are $\frac{\partial f}{\partial x}$ and $\frac{\partial f}{\partial y}$ (Rmk. 5.13), and they are defined as f_x and f_y respectively.

Remark 5.13. The notation $\frac{\partial f}{\partial x}$, is used in functions with two or more variables $f : \mathbb{R}^2 \to \mathbb{R}$ (Def. 5.7), whilst the notation $\frac{df}{dx}$ is used only in functions of one f: $\mathbb{R} \to \mathbb{R}$.

Example 5.35. Let the function $f(x, y) = xy^3$. (i) Determine $\frac{\partial f}{\partial x}$. (ii) $\frac{\partial f}{\partial y}$.

Solution 5.35. (i) $\frac{\partial f}{\partial x} = [xy^3]'_x = x[y^3]'_x + [x]'_x y^3 = 0 + 1y^3 = y^3$. (ii) $\frac{\partial f}{\partial y} = [xy^3]'_y = x[y^3]'_y + [x]'_y y^3 = 3xy^2 + 0 = 3xy^2$.

Remark 5.14. $x[y^3]'_x = 0$, if y is a constant and its derivative is zero, then $[x]'_y = 0$ x, in this case x is a constant.

Theorem 5.4. Let an implicit function $G : \mathbb{R}^2 \to \mathbb{R}$ (Def. 7) and a point $(x_0, y_0) \in D_G$. Determine if (i) its $\frac{\partial G}{\partial x}$ and $\frac{\partial G}{\partial y}$ are C^1 class, (ii) $G(x_0, y_0) = 0$, and (iii) $\frac{\partial G}{\partial y} \neq 0$

then exists as subset I, where $x_0 \in I$, and a unique function f C^1 class such that $f: I \to \mathbb{R}$, which solves for G. Thus $G(x, f(x)) = 0, \forall x \in I$ and $\dfrac{df}{dx} = -\dfrac{\frac{\partial G}{\partial x}}{\frac{\partial G}{\partial y}}$.

Proof. Proof is out of the scope of this book, however, it is stated with an example to show its importance, for further reading [29].

Example 5.36. Be the equation $x^2 + y^3 - 1 = 0$. (i) Obtain $\dfrac{dy}{dx}$ using the Implicit Function Theorem (Thm. 5.7).

Solution 5.36.

$$
\begin{aligned}
f(x, y) &= x^2 + y^3 - 1 \\
\frac{\partial f}{\partial x} &= 2x \\
\frac{\partial f}{\partial y} &= 3y^2 \\
\frac{dy}{dx} &= -\frac{\frac{\partial f}{\partial x}}{\frac{\partial f}{\partial y}} = -\frac{2x}{3y^2}.
\end{aligned}
\tag{5.9}
$$

$\dfrac{dy}{dx}$ exists when $y \neq 0$, then the domain of f is $\mathbb{R} - \{-1,1\}$.

Example 5.37. Be the equation $\sin x + \cos y = e^x$. (i) Obtain $\dfrac{dy}{dx}$ using the Implicit Function Theorem (Thm. 5.7). (ii) Determine the domain of f.

Solution 5.37. (i)

$$
\begin{aligned}
f(x, y) &= \sin x + \cos y - e^x \\
\frac{\partial f}{\partial x} &= \cos x - e^x \\
\frac{\partial f}{\partial y} &= -\sin y \\
\frac{dy}{dx} &= -\frac{\frac{\partial f}{\partial x}}{\frac{\partial f}{\partial y}} = \frac{\cos x - e^x}{\sin y}.
\end{aligned}
\tag{5.10}
$$

(ii) $D_f = \{x \in \mathbb{R} \mid x \neq n\pi, \ n \in \mathbb{Z}\}$.

Example 5.38. Let the equation $\sin t + \cos t = e^x$. (i) Obtain $\dfrac{dT}{dt}$ using the Implicit Function Theorem (Thm. 5.7). (ii) Determine the domain of the map T.

Solution 5.38. (i) Let $y = T(t)$

$$
\begin{aligned}
G(x,y) &= \sin t + \cos y - e^t \\
\frac{\partial G}{\partial t} &= \cos t - e^t \\
\frac{\partial G}{\partial y} &= -\sin y \\
\frac{dy}{dt} &= -\frac{\frac{\partial G}{\partial t}}{\frac{\partial G}{\partial y}} = \frac{\cos t - e^t}{\sin y} \\
\frac{dT}{dt} &= \frac{\cos t - e^t}{\sin T(t)}
\end{aligned} \tag{5.11}
$$

(ii) $D_f = \{t \in \mathbb{R} \mid t \neq n\pi, \ n \in \mathbb{Z}\}$.

An important application of this theorem is for knowing the subset of the domain of function f, where its derivative f' exists, and to know the derivative itself. This implies that this particular subset of the function is a **bijection**, and therefore, its inverse function f^{-1} exists.

Remark 5.15. This theorem cannot be used to know either function f or its inverse function f^{-1}.

5.8. INVERSE FUNCTION THEOREM

This theorem can be used to determine the derivative of an inverse function $(f^{-1})'$, from the derivative of function f. Its application will be reviewed in Taylor's series (Ch. 7). In order to show how it works, we introduced a variation of the map T (Sect. 2.2). This section applies for functions or maps.

Theorem 5.5. Let a function f of C^1 class and a point $x_0 \in D_f$ where $f'(x_0) \neq 0$, then there exists a function $f^{-1}(y_0)$ where $y = f(x_0)$ such that $(f^{-1}(y_0))' = \frac{1}{f'(x_0)}$.

Proof. Proof is out of the scope of this book, however, it is stated with an example for its relevance, for further reading [29].

Example 5.39. Let the function $f(x) = xe^x$ and its equivalent map $T(t) = te^x$. (i) Compute $(f^{-1})'$ at point $x_0 = 1$ using the Inverse Function Theorem (Thm. 5.8).

(ii) Compute $(T^{-1})'$ at point $t_0 = 1$ using the Inverse Function Theorem (Thm. 5.8).

Solution 5.39. (i) Since function f is C^1 class and $f'(x_0) = 2e \neq 0$, then $(f^{-1}(y_0))' = \frac{1}{f'(x_0)} = \frac{1}{e^{x_0} + xe^{x_0}} = \frac{1}{2e}$. (ii) Since map T is C^1 class and $T'(t_0) = (2e) \neq 0$, then $(T^{-1}(y_0))' = (\frac{1}{T'(t_0)}) = (\frac{1}{e^{t_0} + te^{t_0}}) = (\frac{1}{2e})$.

This theorem is useful to obtain the derivative of the inverse function $(f^{-1})'$, without knowing the inverse function of f^{-1}.

Remark 5.16. This theorem assumes that the function f is known.

5.9. THE IMPLICIT AND THE INVERSE FUNCTION THEOREMS

When these theorems are reviewed for the first time, we get the impression that something is missing. The Implicit Function Theorem gives the conditions to explicitly know the derivative f' of a function f, it can even give us the subset of the domain of function f where its inverse f^{-1} exists, however, we don't know function f. On the other hand, the Inverse Function Theorem is a tool to know the derivative of the inverse function $(f^{-1})'$ from the derivative f' of function f, but it doesn't let us know function f, or its inverse function f^{-1}.

In summary, both theorems let us know explicitly the derivatives f' of function f and of its inverse $(f^{-1})'$, but they don't let us know the functions f or f^{-1}. When we review Taylor's series (Ch. 7), we will find that from the derivative f' of a function f, we can determine function f, the same when using the derivative of the inverse function $(f^{-1})'$ we can determine the function f^{-1}.

These theorems and Taylor's series make possible to connect two spaces from a subset that is a **bijection**. First knowing the derivatives and then, from Taylor's series, knowing the functions. The Implicit and Inverse Function Theorems are fundamental in Differential Calculus.

5.10. DERIVATIVE OF A RECIPROCAL THEOREM

Theorem 5.6. Be the function f differentiable at a point x and $f(x) \neq 0$, then $h(x) = 1/f(x)$ is also differentiable at x and $h'(x) = \frac{d}{dx}\frac{1}{f(x)} = \frac{f'(x)}{f^2(x)}$.

Proof. If f is C^1 class and $f(x) \neq 0$, then $g'(x) = \dfrac{d}{dx}\dfrac{1}{f(x)} = \lim_{h \to 0}\left(\dfrac{\frac{1}{f(x+h)} - \frac{1}{f(x)}}{h}\right) = (\lim_{h \to 0}\dfrac{f(x+h)-f(x)}{h})(\lim_{h \to 0}\dfrac{1}{f(x)f(x+h)}) = -f'(x)\dfrac{1}{f^2(x)}$.

Example 5.40. Compute the derivative of the function $f(x) = \sec x$ [68].

Solution 5.40. $(\sec x)' = (\dfrac{1}{\cos x})' = \dfrac{-(\cos x)\prime}{\cos x^2} = \dfrac{\sin x}{\cos x^2} = \sec x \tan x$.

Example 5.41. Compute the derivative of the map $T(t) = \csc x$.

Solution 5.41. $(\csc t)' = (\dfrac{1}{\sin t})' = \dfrac{-(\sin t)\prime}{\sin t^2} = -\dfrac{\cos t}{\sin t^2} = -\cos t \cot t$.

5.11. CASE STUDY: MAXIMIZING VOLUME

Case 5.1. Suppose we want to build a rectangular box of $30\text{cm} \times 40\text{cm}$, leaving a square edge x at the ends to fold. What will be the value of x to have the maximum volume?

First, we have to determine the function of the volume $V(x)$ in terms of the variable x. Since it is a parallelepiped of base $(30 - 2x) \times (40 - 2x)$ and height x, the function of the volume is $V(x) = x(30 - 2x)(40 - 2x) = 4x^3 - 140x^2 + 1200x$ (Fig. **5.4**).

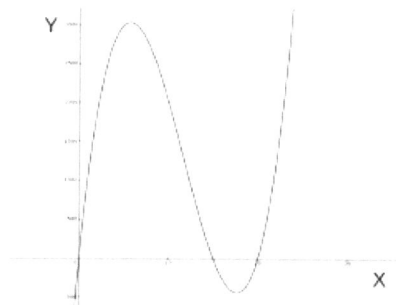

Fig. (5.4). Graph of function $V(x)$.

Now, we will determine the derivative $V'(x) = 12x^2 - 280x + 1200$. What are the critical points? (Def. 5.3). We obtain these points with $V'(x) = 0$, which is equivalent to $12x^2 - 280x + 1200 = 0 \Leftarrow 3x^2 - 70x + 300 = 0$. The values

$x = 5.65$ and $x = 17.84$ solve the equation, so $x = 5.65$ and $x = 17.84$ are the critical points.

Let us apply now the derivative test (Def. 5.3). For this purpose, we require the second derivative of V which is $V''(x) = 24x - 280$, then $V''(5.65) = -144 < 0$ and $V''(17.84) = 148 > 0$. From the derivative test, function V reaches its maximum value when $V''(x) < 0$, which is at point $x = 5.65$ (Fig. **5.4**).

Therefore, the maximum volume is obtained with a base of 18.7×28.7 and a height of 5.65, so this volume is $3032m^3$.

5.12. CASE STUDY: PERIMETER FENCE

Case 5.2. We want to build a perimeter fence for a vegetable garden with a wire mesh and a boundary wall (Fig. **5.5**), The vegetable garden will have width w and length l. We only have 100 meters of wire mesh. What will be the width and length to get a maximum surface? (Case taken and adapted from [13]).

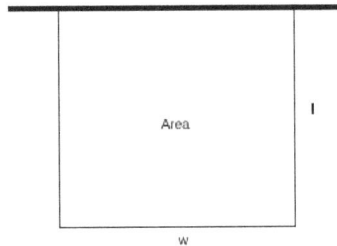

Fig. (5.5). Graph of the area A.

The $Area = w \times l$.

The wire mesh will be required only at 3 sides and these sides must total 100 metres. Then $w = 100 - l - l = 100 - 2l$.

So the $Area(l) = (100 - 2l)l = 100l - 2l^2$.

Now we have to determine its derivative $A'(l) = 100 - 4l$. What are the critical points? (Def. 5.3). We have $A'(l) = 0$, which is equivalent to $100 - 4l = 0$, therefore, there is only one critical point $l = 25$.

To apply the Derivative Test (Def. 5.3), we need the second derivative of A, which is $A''(l) = -4$. So, $A''(25) = -4 < 0$. From the derivative test, function A gets the maximum value when $A''(l) < 0$ at the point $l = 25$.

Therefore, the length of the garden is 25 meters and the width is 50 meters.

5.13. CASE STUDY: HOW FAST IS THE WATER LEVEL RISING?

Case 5.3. Suppose we pour water into a conical container at the rate of $20 cm^3/sec$. The cone points directly down and it has a height of $45cm$ and a base radius of $15cm$ (Fig. **5.6**). How fast is the water level rising when the water is $8cm$ deep, (at its deepest point)?

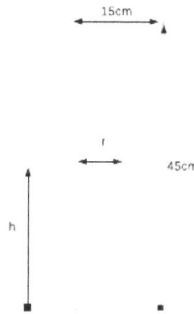

Fig. (5.6). Model of water level rising.

There are three factors that vary in time: the level of the water h, the radius of the surface r, and the volume of the water V.

The volumen V of the cone is given by $V = \frac{\pi r^2 h}{3}$, we can determine $\frac{dV}{dt}$, we want to know $\frac{dh}{dt}$ but we don't know $\frac{dr}{dt}$. However, the dimension of the cone of water must be the same of the container. That is because of similar triangles, $\frac{r}{h} = \frac{15}{45}$, sor $= \frac{h}{3}$. Substituting r in the volumen formula the cone is

$$V = \frac{\pi(\frac{h}{3})^2 h}{3} = \frac{\pi h^3}{27}.$$

Now, let's determine the derivative of V, substituting $h = 8$ and $\frac{dV}{dt} = 20$ we will get $20 = 3\pi \frac{8^2}{27} \frac{dh}{dt}$. Thus $\frac{dh}{dt} = \frac{90}{64\pi} = 0.4476 cm/sec$.

5.14. EXERCISES

Exercise 5.1. Using the definition of limit. (i) Determine the derivative of the function $f(x) = x^2$ at point 3. (ii) Determine the line of the graph of the function at that point. (iii) Obtain the graph of the tangent line and the function f. (iv) Compute the derivative of f using the definition (Def. 5.1.4).

Exercise 5.2. From function $x^2 - 2x + 3$. (i) Determine the critical points. (ii) What are they maximum or minimum?

Exercise 5.3. The difference between two numbers is 10. What is the minimum product?

Exercise 5.4. Let $f(x) = x^3 - 3x + 1$. Find the values $c \in (1,3)$, such that,

$$f'(c) = \frac{f(3) - f(1)}{3 - 1}.$$

Exercise 5.5. Minimize the metalic surface of a closed cylinder with h height and r radious, with capacity of $330ml$.

Exercise 5.6. Let the map $T(t) = (\cos t, \sin t)$, where $t \in [0, 2\pi]$. (i) Using the definition (Def. 5.1.4), determine the derivative of the map T at point π. (ii) Determine the tangent line to the graph of the map, at that point. (iii) Obtain the graph of the tangent line and the map T.

Exercise 5.7. Compute $\lim_{x \to 3} \frac{1}{x-3} - \frac{5}{x^2-x-6}$ using the L'Hospital's Rule [69].

Exercise 5.8. Obtain the derivative $\frac{dy}{dx}$ of [70]: (i) $6x - 2y = 0$. (ii) $\sec^2 x + \csc^2 x$.

Exercise 5.9. Without explicitly obtaining the functions (Ex. 5.7), determine the domain of f when f is a bijection.

Exercise 5.10. Show that $(x^n)' = nx^{n-1}, \forall n \in \mathbb{R}, \text{and} x > 0$ [71].

CHAPTER 6

Sequences

Abstarct: This chapter defines the concept of **arithmetic sequences** or **sequences**. Here, we will review different types of sequences, their properties, limits, and convergence, as well as the important Cauchy sequences, we will also see an extension of them whose elements are functions, the Sequences of functions.

Keywords: Arithmetic sequences, Bounded and unbounded sequences, Cauchy sequences, Convergent and divergent sequences, Graph of sequence (a_n), Infinite and finite sequences, Properties of limit of sequences, Subsequences, The boundedness of convergent sequences theorem.

6.1. ARITHMETIC SEQUENCES

In general, a **sequence** is an enumerated collection of objects in which repetitions are allowed [72]. For instance, $\{C, A, T\}$ is a sequence of letters. In this chapter, we will focus on the **arithmetic sequences** *e.g.* $\{2, -1, 5, 6, 5, 0\}$, or $\{.9, .99, .999, \cdots\}$, and they will be called **sequences**.

Remark 6.1. Two sets are equal if, and only if, they have the same elements. For instance $\{0, 4, 5\} = \{4, 5, 0\}$ here, the order of the elements does not matter; and $\{1, -1\} = \{1, -1, 1, -1, 1\}$ here, the duplication of the elements do not make any difference for the sets [73].

Definition 6.1. A sequence is a function $f: A \subset \mathbb{N} \to S$. Each element $n \in A$ $f(a)$ is denoted a_n and f itself is mostly denoted (a_n) [74]. The **domain** of the sequence (a_n) is formed by the numbers $n \in \mathbb{N}$. The **image** of the sequence (a_n) is formed by the numbers $a_n \in \mathbb{R}$.

6.1.1. Graph of Sequence (a_n)

The graph of a sequence (a_n) on the real line $\mathbb{R}$ is formed by a collection of points $a_1, a_2, \cdots, a_n$, while in the plane $\mathbb{R}^2$ the points (x, y) are formed by (n, a_n) (Fig. **6.1**).

Carlos Polanco
All rights reserved-© 2020 Bentham Science Publishers

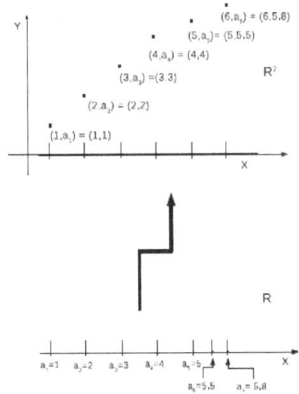

Fig. (6.1). Graph of the sequence (a_n) on $\mathbb{R}$ and its equivalent on $\mathbb{R}^2$.

6.1.2. Infinite and Finite Sequences

The **length** of a sequence is defined as the number of terms a_n in the sequence (a_n).

Definition 6.2. A **finite** sequence (a_n) is a sequence whose **image** is formed by a **finite** number of elements, otherwise it is an **infinite** sequence.

Remark 6.2. A sequence is defined as **infinite** or **finite**, depending on the length of its image.

Example 6.1. Classify the sequences. (i) $(a_1) = \{4,5,6,10\}$. (ii) $(a_2) = \{2,4,6,\cdots\}$. (iii) $(a_3) = \frac{1}{n}$. (iv) $(a_4) = (-1)^n$ (Fig. **6.2**) (v) $(a_5) = \frac{1}{n}$ where its domain is $[5,10] \subset \mathbb{N}$. (vi) $(a_n) = 4$.

Solution 6.1. (i) It is a **finite** sequence because the number of elements forming its image is four. (ii) It is an **infinite** sequence because its image is formed by an infinite number of elements. (iii) It is an **infinite** sequence because its image, although bounded by the interval $[0,1]$, is formed by an infinite number of elements. (iv) It is a **finite** sequence because its image is formed by two elements 1 and -1. (v) It is a **finite** sequence because its image is formed by six elements, as its domain only has six elements. (vi) It is a **finite** sequence because its image is formed by a unique element.

6.1.3. Bounded and Unbounded Sequences

Definition 6.3. The sequence (a_n) is a sequence **bounded from above** if there exists $M \in \mathbb{R}$ such that $a_n \leq M$, $\forall n \in \mathbb{N}$. A sequence is **bounded from below** if there exists $m \in \mathbb{R}$ such that $m \leq a_n, \forall n \in \mathbb{N}$. A sequence is called **bounded** if it is both bounded from above and below [75], otherwise it is called an **unbounded** sequence.

Remark 6.3. It is important to note that the value of M is located in the **image** of the sequence (a_n).

Example 6.2. Classify the sequences. (i) $(a_n) = \frac{1}{n}$. (ii) $(a_n) = (-1)^n$ (Fig. **6.2**). (iii) $(a_n) = \frac{1}{n}$ whose domain is $[5,10] \subset \mathbb{N}$.

Solution 6.2. (i) It is an **infinite** and **bounded from below** sequence, its image is the interval $[0,1]$. (ii) It is an **infinite** and **bounded from below** sequence, its image is bounded in the interval $[-1,1]$ (Fig. **6.2**). (iii) It is a **finite** and **bounded from below** sequence, its image is formed by the interval $\frac{1}{10} \frac{1}{5}$.

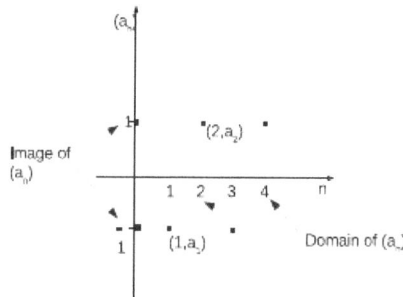

Fig. (6.2). Graph of the sequence $(a_n) = (-1)^n$ on $\mathbb{R}^2$.

Remark 6.4. Two sets are equal if, and only if, they have the same elements. For instance, $\{0,4,5\} = \{4,5,0\}$ the order of the elements does not matter and $\{1, -1\} = \{1, -1, 1, -1, 1\}$ the duplication of the elements does not make any difference for he sets [73].

Definition 6.4. A sequence is **strictly monotonically increasing** if $a_1 < a_2 < a_{n-1} < a_n < a_{n+1} < \cdots$ or $\forall n \in \mathbb{N} \ a_{n+1} > a_n$. On the contrary, a sequence is **strictly monotonically decreasing** if $a_1 > a_2 > a_{n-1} > a_n > a_{n+1} > \cdots$ or $\forall n \in \mathbb{N} \ a_{n+1} < a_n$.

6.1.4. Convergent and Divergent Sequences

Definition 6.5. A **sequence** (a_n) **converges** in a **unique** value $L \in \mathbb{R}$ if $\forall \varepsilon > 0$ exists a value $n \in \mathbb{N}$ such that $\forall n \geq N$, thus we have $|a_n - L| < \varepsilon$ [76]; otherwise the sequence **diverges**. So we can say the sequence is a **convergent** sequence or a **divergent** sequence.

Note that the **direction** of the convergence of the elements a_n, when approaching a limit value, tends to $n \to \infty$ tends to.

Remark 6.5. A sequence (a_n) is convergent if exists a value $L \in \mathbb{R}$ such that for every $\varepsilon > 0$ exists a positive integer N, depending on ε, that $|a_n - L| < \varepsilon$ for all $n \geq N$; i.e. $\lim_{n \to \infty} a_n = L$. The point L is an **accumulation point** or **cluster point** of the sequence (a_n).. The accumulation points belong to the **image** of the sequence.

Remark 6.6. The **unicity** of the **accumulation point** in a **convergent sequence** is given by the **unicity** of the **limit** of the sequence.

Remark 6.7. If any neighbourhood of point L has all the elements of the sequence, except a finite number of them, then the limit of the sequence when $n \to \infty$ is the point L [77]

Proposition 6.1. A bounded sequence (a_n) with a unique accumulation point is convergent.

Proof. From (Rmk. 6.5), the accumulation point is the unique number where the sequence converges, therefore, (a_n) is a convergent sequence.

Remark 6.8. A sequence with two accumulation points is an **Oscillating sequence**.

Example 6.3. Classify the sequence (i) $(a_n) = (-1)^n$ and $(a_n) = \frac{1}{n}$.

Solution 6.3. (i) It is an infinite, bounded, and divergent sequence with two accumulation points -1 and 1. It is bounded in the interval $[-1,1]$. (ii) It is an infinite, bounded, and convergent sequence with the accumulation point 0. It is bounded in $[0,1]$.

6.2. PROPERTIES OF LIMIT OF A SEQUENCES

If $(a_n) \to a$ and $(b_n) \to b$ are **convergent sequences** (Def. 6.5), then [72]

Property 1. $\lim_{n\to\infty} a_n + b_n = \lim_{n\to\infty} a_n + \lim_{n\to\infty} b_n = a + b$ (addition).
Property 2. $\lim_{n\to\infty} a_n - b_n = \lim_{n\to\infty} a_n - \lim_{n\to\infty} b_n$ (substration).
Property 3. $\lim_{n\to\infty}(a_n)(b_n) = (\lim_{n\to\infty} a_n)(\lim_{n\to\infty} b_n)$ (multiplication).
Property 4. $\lim_{n to\infty} \frac{a_n}{b_n} = \frac{\lim_{n\to\infty} a_n}{\lim_{n\to\infty} b_n}$, $\lim_{n\to\infty} b_n \neq 0$, $\forall n \in \mathbb{N}$ (division).

If $a_n < b_n$ for some N, then $\lim_{n\to\infty} a_n < \lim_{n\to\infty} b_n$, therefore, if the sequence (c_n) meets $(a_n) \leq (c_n) \leq (b_n)$, $\forall n > N \in \mathbb{N}$ and $\lim_{n\to\infty}(a_n) = L = \lim_{n\to\infty}(b_n)$, then $\lim_{n\to\infty}(c_n) = L$ and the sequence (c_n) is a convergent sequence (Def. 6.5).

Theorem 6.1. Suppose that (a_n), (b_n), and (c_n) are sequences such that $b_n \leq a_n \leq c_n$ for all n $\lim_{n\to\infty} b_n = L$ and $\lim_{n\to\infty} c_n = L$, so $\lim_{n\to\infty} a_n = L$.

Proof. If $\lim_{n\to\infty}(a_n) = L$ and $\lim_{n\to\infty}(b_n) = L$, then $L \leq \lim_{n\to\infty}(a_n) \leq L$. Therefore, the limit of (a_n) is L [78].

Example 6.4. Let the sequences $(a_n) = 2 + \frac{1}{\sqrt{n}}$ and $(b_n) = 1 - \frac{1}{n}$. Verify (Prop. 6.2)

Solution 6.4. (i) $\lim_{n\to\infty} a_n + b_n = \lim_{n\to\infty} 2 + \frac{1}{\sqrt{n}} + \lim_{n\to\infty} 1 - \frac{1}{n} = 2 - 1 = 1$.

(ii) $\lim_{n\to\infty} a_n - b_n = \lim_{n\to\infty} 2 + \frac{1}{\sqrt{n}} - \lim_{n\to\infty} 1 - \frac{1}{n} = 2 + 1 = 3$.   (iii)

$\lim_{n\to\infty} a_n b_n = (\lim_{n\to\infty} 2 + \frac{1}{\sqrt{n}})(\lim_{n\to\infty} 1 - \frac{1}{n}) = (2)(1) = 2$. $(iv) \lim_{n\to\infty} \frac{a_n}{b_n} = (\lim_{n\to\infty} 2 + \frac{1}{\sqrt{n}})/(\lim_{n\to\infty} 1 - \frac{1}{n}) = \frac{2}{1} = 2$.

6.3. THE BOUNDEDNESS OF CONVERGENT SEQUENCES THEOREM

Theorem 6.2. If a sequence (a_n) is convergent, then (a_n) is bounded.

Proof. If the sequence (a_n) is convergent, then exists a value $N \in \mathbb{N}$ such that $\forall n \geq N$, we have $|a_n - L| < \varepsilon_0$ (Def. 1.4). Equivalently, $-\varepsilon < a_n - L < \varepsilon$ and so $L - \varepsilon < a_n < L + \varepsilon$ for $n \geq N$. Now there is only a finite number of terms to consider, namely the terms $a_1, a_2, \cdots, a_{N-1}$. Let $M = \max\{|a_1|, |a_2|, \cdots, |aM - 1|, |L| + \varepsilon\}$. Therefore, for all $n > N$, $|a_n| < M$ and so (a_n) is bounded.

Remark 6.9. The converse **is not** true. For instance, the sequence $(a_n) = (-1)^n$ (Fig. **6.2**) is a bounded sequence but it is **not** convergent [78].

Example 6.5. Be the convergent sequence $(a_n) = \frac{1}{n}$. What is its bound?

Solution 6.5. Its bound is $[0,1]$.

6.4. CAUCHY SEQUENCES

Definition 6.6. A sequence (a_n) is a Cauchy sequence, if for every $\varepsilon > 0$ there is an $N > 0$ such that $m, n > N \Rightarrow |a_n - a_m| < \varepsilon$ [80].

Proposition 6.2. A Cauchy sequence is a convergent sequence.

Proof. From (Prop. 6.1), if a Cauchy sequence is bounded and it has a unique accumulation point, then it is convergent.

Example 6.6. Let the sequence $(a_n) = -\frac{1}{n}$. Show that this sequence is a Cauchy sequence.

Solution 6.6. The sequence (a_n) is a bounded sequence in the interval $[0,1]$ and it has a unique accumulation point 0. This sequence is a Cauchy sequence.

6.5. SUB SEQUENCES

Definition 6.7. A **sub sequence** $(a_n)_i$ of a given sequence (a_n) is a **sequence** formed by omitting some of the elements of the sequence, without affecting the relative position of the original elements.

Example 6.7. Let the sequence $(a_n) = 1,2,3,4,\cdots, n$. Show two sub sequences of (a_n).

Solution 6.7 $(a_n)_i = \{2,4,6,\cdots,2n\}$, $n \in \mathbb{N}$. $(a_n)_j = \{1,3,5,\cdots,2n + 1\}$, $n \in \mathbb{N}$.

6.6. CASE STUDY: TERM $\frac{5n+1}{n}$

Case 6.1. (i) Proof that the sequence (a_n), whose general term is $\frac{5n+1}{n}$, has limit 5. (ii) Find the terms of the sequence that are outside the neighbourhood $(5 - 0.001, 5 + 0.001)$.

(i) The $\lim_{n\to\infty} \frac{5n+1}{n} = \lim_{n\to\infty} \frac{\frac{5n+1}{n}}{\frac{n}{n}} = \lim_{n\to\infty} \frac{5+\frac{1}{n}}{1} = \lim_{n\to\infty} 5 = 5.$

(ii) In order to find the terms that are outside the neighbourhood, we solve:

$$\left|\frac{5n+1}{n} - 5\right| < \frac{1}{1000} \Leftrightarrow \left|\frac{5n+1-5n}{n}\right| < \frac{1}{1000} \Leftrightarrow \left|\frac{1}{n}\right| < \frac{1}{1000} \Leftrightarrow \frac{1}{n} < \frac{1}{1000}.$$

This implies that $n < 1000$, so the convergence starts at 1001.

6.7. CASE STUDY: TERM $\frac{n+1}{n+2}$

Case 6.2. To proof that the sequence (a_n) is divergent, we have to show that it is not a Cauchy sequence, and therefore, it is not convergent (Prop. 6.2).

Proof. If the sequence is **convergent** it must meet the condition (Def. 4) $|a_n - a_m| < \varepsilon$. Taking $\varepsilon = 0.2$, $a_{n=1} = \frac{(1)+1}{(1)+2} = \frac{2}{3}$, and $a_{n=20} = \frac{(20)+1}{(20)+2} = \frac{21}{22}$; substituting these values in the inequality $|a_n - a_m| = \left|\frac{2}{3} - \frac{21}{22}\right| < \varepsilon = 0.2 \Leftarrow |-0.28| = 0.28 < 0.2!$. So the sequence is divergent.

6.8. CASE STUDY: TERM $\frac{a_{n-2}+a_{n-1}}{2}$, $n > 3$

Let $a_1 = 1$ and $a_2 = 3$. Proof that the sequence (a_n) is convergent (Case taken and adapted from [17]).

Proof. If the sequence is **convergent** it must meet the condition (Def. 4) $|a_n - a_m| < \varepsilon$. If $a_n = \frac{a_{n-2}+a_{n-1}}{2}$, then $a_{n+1} = \frac{a_{n-1}+a_n}{2}$. This implies $\left|\frac{a_n-a_{n-2}}{2}\right| < \varepsilon$, which is equivalent to $\left|\frac{a_{n-1}}{4}\right| < \varepsilon$ and also equivalent to $\left|\frac{a_{n-3}+a_{n-2}}{8}\right| < \varepsilon$. Note that $\lim_{n\to\infty} \frac{a_{n-3}+a_{n-2}}{2^{n-3}} = 0$, therefore, the sequence is convergent.

6.9. EXERCISES

Exercise 6.1. Give a sequence. (i) Monotonically and not bounded. (ii) Bounded not monotonically. (iii) Non-bounded and non-monotonically. (iv) Non-bounded and convergent. (v) Bounded and divergent. (vi) Bounded and non-convergent. (vii) Non-monotonically and convergent. (viii) Non-monotonically and divergent.

Exercise 6.2. Compute the **general term** of the sequence $8, 3, -2, -7, -12, \cdots$.

Exercise 6.3. Find the sequence whose general term is $a_n = (1 + \frac{x}{n})^n$.

Exercise 6.4. Find the general term of the sequence $\{0, 3, 8, 15, 24, 35\}$.

Exercise 6.5. Is the sequence $\{1, 2, 3, 4, \cdots\}$ a convergent sequence?

Exercise 6.6. Give the definition of the **recursive sequence** and give an example of it.

Exercise 6.7. Explain the difference of a representation of a sequence in a plain and in a real line.

Exercise 6.8. Is the value 25 a term of the sequence whose general term is $a_n = 3n + 12$.

Exercise 6.9. Describe the sequence $(a_n) = (-1)^{n-1} 2^n$.

Exercise 6.10. Let the sequence $(a_n) = \frac{1}{n}$. What can you say about the subsequence $\frac{1}{2k}$, where $k \in \mathbb{N}$?

CHAPTER 7

Series

Abstarct: This chapter focuses on the definition of convergence, its classification and the arithmetic series, particularly Fourier and Taylor series.

Keywords: Alternating series, Binomial series, General convergence criteria, Maclaurin series, Particular convergence criteria, Power series, Properties of series, Sequence of partial sums, Series of positive terms, Taylor series, Telescoping series.

7.1. PRELIMINARIES

In this section, we define series and their characteristics, we review the main types of series by their usefulness such as Taylor series and Maclaurin series. We particularly see how the limit of a series determines the area of the graph of a function in a closed interval.

7.2. DEFINITION

Definition 7.1. A series S is the **sum** of n terms of a **sequence** (a_i)

$$S = \sum_{i=1}^{n} a_i = a_1 + a_2 + a_3 + \cdots + a_n$$

A series S is **finite** $S = \sum_{i=1}^{n<\infty} a_i = a_1 + a_2 + \cdots + a_n$ or **infinite** $S = \sum_{i=1}^{n=\infty} a_i = a_1 + a_2 + \cdots$ depending if n represents a **finite** or an **infinite** value.

If the **limit** of the **series** meets $\lim_{n\to\infty} \sum_{i=1}^{n} a_i = L \in \mathbb{R}$, then it is a **convergent series**, otherwise it is a **divergent series**.

7.3. SEQUENCE OF PARTIAL SUMS

Given a series $\sum_{i=1}^{k} a_i$, it is possible to have a **sequence** (a_S) with the elements of series S, *i.e.* $a_S = \{a_{S=1}, a_{S=2}, a_{S=3}, \cdots\} = \{\sum_{i=1}^{1} a_i, \sum_{i=1}^{2} a_i, \sum_{i=1}^{3} a_i, \cdots\}$. The sequence (a_S) is called **sequence of partial sums** and (a_S) is an **increasing sequence** since $a_{S+1} > a_S$, otherwise the sequence is a **decreasing sequence**.

Carlos Polanco
All rights reserved-© 2020 Bentham Science Publishers

Remark 7.1. The **sequence of partial sums** (a_S) is **different** than the **sequence** (a_n), whose general term a_i is included in the series $\sum_{i=1}^{k} a_i$.

Example 7.1. Give an example of (i) an **increasing sequence of partial sums**, (ii) a **non-increasing sequence of partial sums**.

Solution 7.1. (i) $\sum_{n=1}^{k} 1$, this series has as **sequence of partial sums** $(a_S) = \{1,2,3,\cdots\}$. (ii) $\sum_{n=1}^{k} (-1)^n$, this series has as **sequence of partial sums** $(a_S) = \{1,0,1,0,\cdots\}$.

7.4. GENERAL CONVERGENCE CRITERIA

In general, a series $\sum_{i=1}^{\infty} a_i$ is **convergent** if $\lim_{n\to\infty} a_S$ **converge**. However, this can be difficult to verify, that is why there are procedures to verify if a series converges or not, depending on the type of series.

Remark 7.2. It is important to note that the methods we are going to review here, help us to know if the series is convergent or not. However, in case it is convergent they **do not** give us the value the series converges to.

Definition 7.2. The series $\sum_{i=1}^{k} a_i$ is **absolutely convergent** if the series $\sum_{i=1}^{k} |a_i|$ is **convergent**.

Example 7.2. Provide an example of (i) a conditionally convergent series, and (ii) an absolutely convergent series.

Sloution 7.2. (i) $\sum_{n=1}^{\infty} (-1)^n \frac{1}{n}$. (ii) $\sum_{n=1}^{\infty} (-1)^n \frac{1}{n^4}$.

Definition 7.3. The series $\sum_{i=1}^{k} a_i$ is **conditionally convergent** if the series $\sum_{i=1}^{k} |a_i|$ is **convergent** but not **absolutely convergent**.

Example 7.3. (i) Is the **sequence** $(a_n) = \frac{1}{n}$ convergent? (ii) Calculate some terms of the **sequence of partial sums** (a_S) of the **series** $\sum_{n=1}^{\infty} \frac{1}{n}$. (iii) Discuss the trend of the terms of the **sequence of partial sums** (a_S). (iv) Determine the **sequence of partial sums** (a_S). (v) Determine $\lim_{n\to\infty} a_S$. (vi) From (v) what can you conclude about the **series** $\sum_{n=1}^{\infty} \frac{1}{n}$?

Solution 7.3. (i) $\lim_{n\to\infty}\frac{1}{n}=0$, the **sequence** is **convergent**. (ii) The first seven terms of the **of partial sums** (a_S) are $\{1,\frac{3}{2},\frac{11}{6},\frac{25}{12},\frac{137}{60},\frac{49}{20},\frac{363}{140},\cdots\}$ in fraction terms or $\{1.00,1.50,1.83,2.08,2.28,2.45,2.59,\cdots\}$. (iii) The first seven terms of the **sequence of partial sums** show that each term is larger than the previous one and that the increment is slow but constant, therefore, it can be gathered that the **sequence of partial sums** is **divergent**. (iv) (**Description taken from** [81])

$$a_{S=2} \quad = 1+\frac{1}{2} \tag{7.1}$$

$$a_{S=4} \quad = 1+\frac{1}{2}+\frac{1}{3}+\frac{1}{4} > 1+\frac{1}{2}+\frac{1}{4}+\frac{1}{4} = 1+\frac{1}{2}+\frac{1}{2} = 1+\frac{2}{2}\,(Rmk.\,7.3)$$

$$a_{S=8} \quad = a_{S=4}+\frac{1}{5}+\frac{1}{6}+\frac{1}{7}+\frac{1}{8} > a_{S=4}+\frac{1}{8}+\frac{1}{8}+\frac{1}{8}+\frac{1}{8} = a_{S=4}+\frac{1}{2} > 1+\frac{3}{2}$$

$$a_{S=16} \quad = a_{S=8}+\frac{1}{9}+\frac{1}{10}+\frac{1}{11}+\frac{1}{12}+\frac{1}{13}+\frac{1}{14}+\frac{1}{15}+\frac{1}{16}$$

$$> a_{S=8}+\frac{1}{16}+\frac{1}{16}+\frac{1}{16}+\frac{1}{16}+\frac{1}{16}+\frac{1}{16}+\frac{1}{16}+\frac{1}{16} > a_{S=8}+\frac{1}{2} > 1+\frac{4}{2}$$

The sequence of partial sums $a_{S=k}=\{1+\frac{1}{2},1+\frac{2}{2},1+\frac{3}{2},\cdots,1+\frac{k}{2}\}$. (v) $\lim_{n\to\infty}a_S = \lim_{n\to\infty}1+\frac{n}{2} = \infty$. (vi) The series $\sum_{n=1}^{\infty}\frac{1}{n}$ is divergent.

Remark 7.3. Convergence or divergence is not affected by the addition or substraction of a finite number at the beginning of a series.

7.5. PARTICULAR CONVERGENCE CRITERIA

Here, we will explain the main convergence criteria for families of series

7.5.1. Series of Positive Terms

Definition 7.4. A **positive term series** is a series whose terms a_i are **positive**.

$$\lim_{n\to\infty}\left|\frac{a_{n+1}}{a_n}\right| = \begin{cases} <1 & : \quad \sum_{n=1}^{\infty}a_n\text{ is absolutely convergent} \\ >1 & : \quad \sum_{n=1}^{\infty}a_n\text{ is divergent} \\ =1 & : \quad \text{is not conclusive} \end{cases} \tag{7.2}$$

Example 7.4. Is the series $\sum_{n=1}^{\infty} \frac{1}{2^n}$ convergent?

Solution 7.4. $\lim_{n\to\infty} \left|\frac{a_{n+1}}{a_n}\right| = \lim_{n\to\infty} \frac{1}{2} = \frac{1}{2}$, the series is convergent.

Example 7.5. Is the series $\sum_{n=1}^{\infty} \frac{1}{n}$ convergent?

Solution 7.5. By (Eq. 7.2) $\lim_{n\to\infty} \left|\frac{a_{n+1}}{a_n}\right| = \lim_{n\to\infty} 1 - \frac{1}{n+1} = 1$, it is not possible to conclude if the series is convergent or divergent. (Ex. 7.2).

$$\lim_{n\to\infty} \left|\sqrt[n]{|a_n|}\right| = \begin{cases} < 1 & : & \sum_{n=1}^{\infty} a_n \text{ is absolutely convergent} \\ > 1 & : & \sum_{n=1}^{\infty} a_n \text{ is divergent} \\ = 1 & : & \text{it is not conclusive} \end{cases} \qquad (7.3)$$

Example 7.6. Is the series $\sum_{n=1}^{\infty} n^n$ convergent?

Solution 7.6. By (Eq. 7.3) $\lim_{n\to\infty} \left|\sqrt[n]{|n^n|}\right| = \lim_{n\to\infty} n = \infty$, the series is divergent.

7.5.2. Alternating Series

Definition 7.5. An **alternating series** is a series whose terms $(-1)^n a_n$ are alternately **positive** and **negative**, *i.e.* $\sum_{n=1}^{\infty} (-1)^n a_n$.

Remark 7.4. Note that the term a_n does not include the term $(-1)^n$.

Theorem 7.1. Let an **alternating series** whose sequence (i) (a_n) is decreasing (Def. 7.3) and (ii) $\lim_{n\to\infty} a_n = 0$ [82], then $\sum_{n=1}^{\infty} (-1)^n a_n$ converges.

Example 7.7. Is the series $\sum_{n=1}^{\infty} (-1)^n \frac{1}{n}$ convergent?

Solution 7.7. Condition (i) $(a_n) = \{1, \frac{1}{2}, \frac{1}{3}, \frac{1}{4}, \cdots\}$ the sequence (a_n) is decreasing and condition (ii) $\lim_{n\to\infty} a_n = \lim_{n\to\infty} \frac{1}{n} = 0$. For (i) and (ii) the series $\sum_{n=1}^{\infty} (-1)^n \frac{1}{n}$ is convergent.

In case the alternating series **does not** meet (Def. 7.5), then use criterion (Eq. 7.2) or (Eq. 7.3).

Example 7.8. Is the series $\sum_{n=1}^{\infty} (-1)^n 2^n$ convergent?

Solution 7.8. The sequence (a_n) is not decreasing (Def. 7.5). By (Eq. 7.2) $\lim_{n\to\infty}$ $|\frac{a_{n+1}}{a_n}| = \lim_{n\to\infty} 2 = 2$, so the series is divergent.

7.5.3. Power Series

Definition 7.6. A series is a **power series** if it has the form

$$\sum_{n=1}^{\infty} (x - c)^n = a_0 + (x - c) + a_1(x - c)^2 + (x - c)^3 + \cdots$$

Note the convergence of the power series $\sum_{n=1}^{\infty} (x - c)^n$ depends on the value of x, *i.e.* $c = 0$ and $x = 10$ the power series $\sum_{n=1}^{\infty} (x)^n = 10 + 10^2 + 10^3 + \cdots$ diverges, when $c = 0$ and $x = 1$ the power series $\sum_{n=1}^{\infty} (x)^n = 1 + 1^2 + 1^3 + \cdots$ diverges. On the other hand, when $c = 0$ and $|x| < 1$ the power series $\sum_{n=1}^{\infty} (x)^n = .9 + .9^2 + .9^3 + \cdots$ converges.

The criteria of (Eq. 7.2) or (Eq. 7.3) help us verify the convergence of the power series.

Example 7.9. Is the power series $\sum_{n=1}^{\infty} x^n$ convergent?

Solution 7.9. By (Eq. 7.2) $\lim_{n\to\infty} |\frac{a_{n+1}}{a_n}| = \lim_{n\to\infty} \frac{|x|^{n+1}}{|x|^n} = \lim_{n\to\infty} |x| = |x|$, if $|x| < 1$, then the series converges.

Remark 7.5. Interval $|x| < 1 \Leftarrow (-1,1)$ is called **radius of convergence** 83.

Example 7.10. Is the power series $\sum_{n=1}^{\infty} 2^n x^n$ convergent?

Solution 7.10. By (Eq. 7.3) $\lim_{n\to\infty} |\sqrt[n]{|n^n|}| = \lim_{n\to\infty} 2|x| = 2|x|$, if $|x| < \frac{1}{2}$, then the series converges.

7.5.4. Telescoping Series

Definition 7.7. A **telescoping series** is a series that holds

$$\sum_{i=1}^{n} a_{i+1} - a_i = a_{n+1} - a_1.$$

Show $\sum_{i=1}^{n} a_{i+1} - a_i = a_{n+1} - a_1$

Proof.

$$
\begin{aligned}
\sum_{i=1}^{n} a_{i+1} - a_i \; &= (a_2 - a_1) + (a_3 - a_2) + (a_4 - a_3) + \cdots + (a_{n+1} - a_n) \\
&= -a_1 + (a_2 - a_2) + (a_3 - a_3) + \cdots + (a_n - a_n) + a_{n+1} \quad \textbf{(7.4)} \\
&= a_{n-1} - a_1
\end{aligned}
$$

Show $\sum_{i=1}^{n} (i+1)^2 - i^2 = (n+1)^2 - 1$

Proof.

$$
\begin{aligned}
\sum_{i=1}^{n} i \; &= (a_2 - a_1) + (a_3 - a_2) + (a_4 - a_3) + (a_{n+1} - a_n) \\
&= (2^2 - 1^2) + (3^2 - 2^2) + (4^2 - 3^2) + \cdots + (n+1)^2 - n^2) \\
&= -1^2(2^2 - 2^2) + (3^2 - 3^2) + \cdots + (n^2 - n^2) + (n+1)^2 \quad \textbf{(7.5)} \\
&= -1 + (n+1)^2 \\
&= (n+1)^2 - 1
\end{aligned}
$$

Show $\sum_{i=1}^{n} i^2 = \frac{n(n+1)}{2}$

Proof.

$$
\begin{aligned}
(i+1)^2 - i^2 \qquad\quad &= 2i + 1 \\
\sum_{i=1}^{n} (i+1)^2 - i^2 \; &= \sum_{i=1}^{n} 2i + 1 \\
(n+1)^2 - 1 \qquad\quad &= 2\sum_{i=1}^{n} i + \sum_{i=1}^{n} 1 \quad \text{(Eq. 7.5)} \\
n^2 + 2n \qquad\qquad\; &= 2\sum_{i=1}^{n} i + n \qquad\qquad\qquad\qquad \textbf{(7.6)} \\
\frac{n(n+1)}{2} \qquad\qquad &= \sum_{i=1}^{n} i
\end{aligned}
$$

Similarly, the condensed formula for the series of the type $\sum_{i=1}^{n} i^k$ can be determined using the identity $(i+1)^{k+1} - i^{k+1}$.

Example 7.11. Is the series $\sum_{i=1}^{n} \frac{1}{i(i+1)}$ (i) a telescoping series? (ii) If it is, find its condensed formula.

Solution 7.11. (i)

$$
\begin{aligned}
\sum_{i=1}^{n} \frac{1}{i(i+1)} &= \sum_{i=1}^{n} \frac{1}{i} - \frac{1}{i+1} \\
&= \left(1 - \frac{1}{2}\right) + \left(\frac{1}{2} - \frac{1}{3}\right) + \left(\frac{1}{3} - \frac{1}{4}\right) + \left(\frac{1}{n} - \frac{1}{n+1}\right) \\
&= 1 - \frac{1}{n+1}
\end{aligned}
\tag{7.7}
$$

then,

$$
\sum_{i=1}^{n} \frac{1}{i(i+1)} = 1 - \frac{1}{n+1}
$$

7.5.5. Properties of Series

Let the **convergent series** $\sum_{i=1}^{n} a_i = A \in \mathbb{R}$ and $\sum_{i=1}^{n} B_i = B \in \mathbb{R}$ then:

Property 1. $\sum_{i=1}^{n} \alpha a_i = \alpha A$ (constant).
Property 2. $\sum_{i=1}^{n} a_i + b_i = A + B$ (addition).
Property 3. $\sum_{i=1}^{n} a_i - b_i = A - B$ (substraction).
Property 4. $(\sum_{i=1}^{n} a_i)(\sum_{i=1}^{n} b_i) = AB$ (multiplication).
Property 5. If $B \neq 0$, $(\sum_{i=1}^{n} a_i)/(\sum_{i=1}^{n} b_i) = \frac{A}{B}$ (division).

Example 7.12. Let the **convergent** series $\sum_{i=1}^{3} a_i = x^3 = 36$ and $\sum_{i=1}^{2} b_i = \frac{1}{n} = \frac{3}{2}$. Compute the properties of the series.

Solution 7.12. (i) $\sum_{i=1}^{n} \alpha a_i = \alpha x^3 = \alpha(1 + 8 + 27) = \alpha 36$. (ii) $\sum_{i=1}^{3} a_i + b_i = 36 + \frac{3}{2} = 37.5$ (iii) $\sum_{i=1}^{3} a_i - b_i = 36 - \frac{3}{2} = 34.5$ (iv) $(\sum_{i=1}^{3} a_i)(\sum_{i=1}^{n} b_i) = (36)(\frac{3}{2}) = 54$. (v) $(\sum_{i=1}^{3} a_i)/(\sum_{i=1}^{3} b_i) = \frac{36}{1.5} = 24$.

7.6. BINOMIAL SERIES

Definition 7.8. Let the function $f(x) = (1+x)^\alpha$, where $\alpha \in \mathbb{R}$ and $\binom{n}{r} = \frac{n!}{(n+r)!r!}$.

$$f(x) = (1+x)^k = \sum_{i=1}^{n} \binom{n}{i} a^{n-i} b^i$$

Example 7.13. What is the binomial series of $f(x) = (2x - 3)^2$?

Solution 7.13.

$$
\begin{aligned}
(2x-3)^2 &= \Sigma_{n=1}^{2} \binom{2}{i} (2x)^{n-i}(-3)^i \\
&= \binom{2}{0}(2x)^2(-3)^0 + \binom{2}{1}(2x)^1(-3)^1 + \binom{2}{2}(2x)^0(-3)^2 \\
&= (1)4x^2 + (2)(2x)(-3) + (1)(9) \\
&= 4x^2 - 12x + 9
\end{aligned}
$$

(7.8)

7.7. TAYLOR SERIES

Definition 7.9. Let the function f be infinitely differentiable at $x = x_0$, then

$$f(x) = \sum_{n=0}^{\infty} \frac{f^{(n)}(x_0)}{n!}(x-x_0)^n = f(x_0) + \frac{f'(x_0)}{1!}(x-x_0) + \frac{f''(x_0)}{2!}(x-x_0)^2 + \cdots$$

Example 7.14. Find the Taylor series for $f(x) = \cos x$ of second degree at the point $x_0 = \pi$?

Solution 7.14.

$$
\begin{aligned}
f(x) &= \Sigma_{n=0}^{2} \frac{f^{(n)}(x_0)}{n!}(x-x_0)^n \\
&= f(x_0) + \frac{f'(x_0)}{1!}(x-x_0) + \frac{f''(x_0)}{2!}(x-x_0)^2 \\
&= \cos x_0 - \frac{\sin x_0}{1}(x-x_0) + \frac{\cos x_0}{2}(x-x_0)^2 \\
&= \cos \pi - \sin \pi (x-\pi) + \frac{\cos \pi}{2}(x-\pi)^2 \\
&= -1 - 0 - \frac{1}{2}(x-\pi)^2 \\
&= -1 - \frac{1}{2}(x^2 - 2\pi x + \pi^2) \\
&= -\frac{1}{2}x^2 + \pi x - \frac{1}{2}\pi^2 - 1
\end{aligned}
$$

(7.9)

Remark 7.6. Note that the last term is a **second degree** polynomial $ax^2 + bx + c$.

7.8. MACLAURIN SERIES

Definition 7.10. Let the function f be infinitely differentiable at $x = x_0$, then

$$f(x) = \sum_{n=0}^{\infty} \frac{f^{(n)}(0)}{n!} x^n = f(0) + \frac{f'(0)}{1!} x + \frac{f''(0)}{2!} x^2 + \frac{f'''(0)}{3!} x^3 + \cdots$$

Remark 7.7. The Maclaurin series is a Taylor series at the point $x_0 = 0$

Example 7.15. Find the Maclaurin series for $f(x) = \cos x$, (i) of second degree and (ii) of n degree

Solution 7.15. (i)

$$\begin{aligned}
f(x) &= \sum_{n=0}^{2} \frac{f^{(n)}(0)}{n!} x^n \\
&= f(0) + \frac{f'(0)}{1!} x + \frac{f''(0)}{2!} x^2 \\
&= \cos 0 - \frac{\sin 0}{1} x + \frac{\cos 0}{2} x^2 \\
&= 1 - 0 + \frac{1}{2} x^2 \\
&= 1 + \frac{1}{2} x^2
\end{aligned} \tag{7.10}$$

Remark 7.8. Note that the last term is a **second degree** polynomial $ax^2 + bx + c$.
(ii)

$$\begin{aligned}
f(x) &= \sum_{n=0}^{\infty} \frac{f^{(n)}(0)}{n!} x^n \\
&= \sum_{n=0}^{\infty} \frac{(-1)^n}{(2n)!} x^{2n}
\end{aligned} \tag{7.11}$$

Example 7.16. Find the Maclaurin series for $f(x) = \sin x$ of n degree

Solution 7.16.

$$\begin{aligned}
f(x) &= \sum_{n=0}^{\infty} \frac{f^{(n)}(0)}{n!} x^n \\
&= \sum_{n=0}^{\infty} (-1)^n \frac{x^{2n+1}}{(2n+1)!}
\end{aligned} \tag{7.12}$$

7.9. CASE STUDY: CALCULATION OF $\sum_{n=1}^{\infty} e^{x^2}$

An approach is to use the Maclaurin series (Def. 7.10) for $f(x) = e^{x^2}$, so we will make an approximation to obtain a fourth degree polynomial $n = 4$.

Let us take the following derivatives of function $\dot{f}$

$$
\begin{aligned}
f'(x) &= 2xe^{x^2} \\
f''(x) &= 4x^2 e^{x^2} + 2e^{x^2} \\
f'''(x) &= (8x^3 + 12x)e^{x^2} \\
f^{iv}(x) &= (16x^4 + 48x^2 + 12)e^{x^2}
\end{aligned}
\tag{7.13}
$$

Now, we will evaluate the function and each derivative in 0

$$
\begin{aligned}
f(0) &= 1 \\
f'(0) &= 0 \\
f''(0) &= 2 \\
f'''(0) &= 0 \\
f^{iv}(x) &= 12
\end{aligned}
\tag{7.14}
$$

Then, we place the values in the Maclaurin series.

$$
\begin{aligned}
f(x) &= \sum_{n=0}^{3} \frac{f^{(n)}(0)}{n!} x^n \\
&= f(0) + \frac{f'(0)}{1!}x + \frac{f''(0)}{2!}x^2 + \frac{f'''(0)}{3!}x^3 + \frac{f^{iv}(0)}{4!} \\
&= 1 + x^2 + \frac{1}{2}x^4
\end{aligned}
\tag{7.15}
$$

7.10. CASE STUDY: LET THE VALUE $e = \sum_{n=0}^{\infty} \frac{1}{n!}$

We can verify this by substituting $x = 1$ in the series $e^x = \sum_{n=0}^{\infty} \frac{x^n}{n!}$. Now, we deduce the latter calculating the Maclaurin series for function $f(x) = e^x$.

7.11. CASE STUDY: AREAS AS LIMIT OF A SERIES

Case 7.1. **Definition 7.11.** Let f be a real-valued function defined in the interval $[a, b]$, such that for all x on the domain of f, $f(x) \geq 0$ (*i.e.*, such that f is positive). The area A under the curve of the graph of function f in this interval and the X-axis is equivalent to

$$A_f = \lim_{n \to \infty} \sum_{i=1}^{n} f(x_i)(x_i - x_{i-1}) = \lim_{n \to \infty} \frac{(b-a)}{n} \sum_{i=1}^{n} f(a + i\frac{(b-a)}{n})$$

Example 7.17. Be the function $f(x) = x$ in the interval $[1,2]$. (i) Draw its graph. (ii) Geometrically compute the area. (iii) Compute its area using Riemann series.

Solution 7.17. (i) (Fig. **7.1**). (ii) The area of the triangle with vertices A, B', and C' is $\frac{1}{2}$. The area of the triangle with vertices A, B, and C is 2. Both triangles are similar, therefore, the area is $2 - \frac{1}{2} = \frac{3}{2}$. (iii)

$$
\begin{aligned}
A_f &= \lim_{n \to \infty} \frac{(b-a)}{n} \sum_{i=1}^{n} f(a + i\frac{(b-a)}{n}) = \lim_{n \to \infty} \frac{(2-1)}{n} \sum_{i=1}^{n} f(1 + i\frac{(2-1)}{n}) \\
&= \lim_{n \to \infty} \frac{1}{n} \sum_{i=1}^{n} f(1 + \frac{i}{n}) = \lim_{n \to \infty} \frac{1}{n} \sum_{i=1}^{n} (1 + \frac{i}{n}) \\
&= \lim_{n \to \infty} \frac{1}{n} [\sum_{i=1}^{n} 1 + \frac{1}{n} \sum_{i=1}^{n} i] = \lim_{n \to \infty} \frac{1}{n} [n + \frac{n^2+n}{2n}] \ (Rmk. \ 5.4) = \lim_{n \to \infty} \frac{3n+1}{2n} = \frac{3}{2}
\end{aligned}
$$

(7.16)

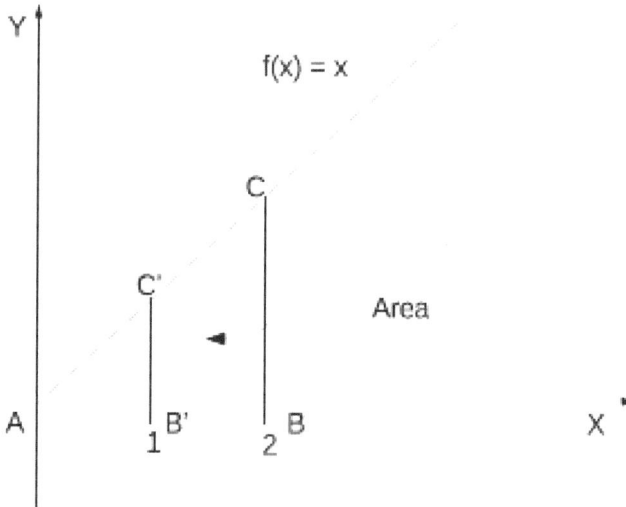

Fig. (7.1). Graph of the function $f(x) = x$.

7.12. EXERCISES

Exercise 7.1. Is the series $\sum_{n=0}^{\infty} \frac{1}{2^n}$ convergent? Explain the meaning of the result.

Exercise 7.2. Is the series $\sum_{n=0}^{\infty} \frac{n!}{n^n}$ convergent?

Exercise 7.3. Describe and give an example of the **convergence criterion of a series with positive terms**.

Exercise 7.4. Describe $\frac{1}{3}$ as **the sum of an infinite series** and determine the value of the series [84].

Exercise 7.5. Is the series $\sum_{n=0}^{\infty} \frac{2n+1}{n^2(n+1)^2}$ convergent? If so, compute its value [84].

Exercise 7.6. Explain how can the Maclaurin series can be used compared with the Taylor series.

Exercise 7.7. What is the relation between the Maclaurin series or Taylor series with the **Implicit Function Theorem** (Sect. 5.7) and the **Inverse Function Theorem** (Sect. 5.8)?

Exercise 7.8. Is the series $\sum_{n=2}^{\infty} \frac{\cos n\pi}{\sqrt{n}}$ convergent? [85].

Exercise 7.9. Is the series $\sum_{n=1}^{\infty} \left(\frac{n^2+1}{2n^2+1}\right)^n$ convergent?

Exercise 7.10. Let the series $\sum_{n=1}^{\infty} \left(\frac{1}{2}\right)^n$. What is its sequence of partial sums? Is the series convergent?

CHAPTER 8

Sequences and Series of Functions

Abstarct: This chapter reviews two operators the **sequence of functions** (f_k) and the **series of functions** $\sum_{k=1}^{n} f_k$. In both cases, we will see the concepts of **uniform** and **pointwise convergence** to the function they converge and the persistence of the properties these functions have with respect to the function f they converge to.

Keywords: Differentiation, Monotonicity, Periodic points, Pointwise convergence, Semi-continuous functions, Sequences of functions, Series of functions, Uniform convergence, Uniform limit theorem.

8.1. PRELIMINARIES

We will define a sequence of functions and study the convergence of this particular type of functions and the function properties they preserve.

8.2. DEFINITION

Definition 8.1. A **sequence of functions** [86] is any countable ordered set of real-valued functions $f_k(x)$, such that $(f_k) = \{f_1(x), f_2(x), \cdots, f_n(x)\}, k \in \mathbb{N}$.

Remark 8.1. The function f does not inherit much of the properties from the sequence of function f_n.

Example 8.1. (i) Define the sequence of function $f_k(x) = x^k$. (ii) Graphically describe the trend of the sequence of function $f_k(x) = x^k$ in the interval $[0,1)$.

Solution 8.1. (i) $(f_k) = \{x, x^2, x^3, x^4, \cdots, x^k\}$.(ii) Fig. (**8.1**) describes the behaviour of the functions f_k in the interval $[0,1)$, it can be seen that the greater the exponent the graphs of f_k tend to the point $(1,0)$.

Determine the behaviour of the sequence of function $f_n(x) = \frac{1}{x^n}$ Fig. (**8.2**) in the interval $(1,2]$ and in the interval $[1,2]$. $(f_n) = x^n = \{\frac{1}{x}, \frac{1}{x^2}, \frac{1}{x^3}, \frac{1}{x^4}, \cdots, \frac{1}{x^n}\}$, the sequence tends from $0 \to 1$ (Eq. 8.1).

Carlos Polanco
All rights reserved-© 2020 Bentham Science Publishers

$$f(x) = \lim_{n \to \infty} f_n(x) = \begin{cases} \infty & for \quad x \in (1,2] \\ 1 & for \quad x \in [1,2] \end{cases} \tag{8.1}$$

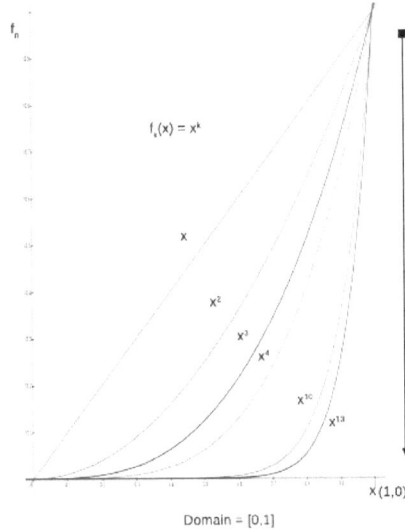

Fig. (8.1). Graph of the function $f_k(x) = x^k$.

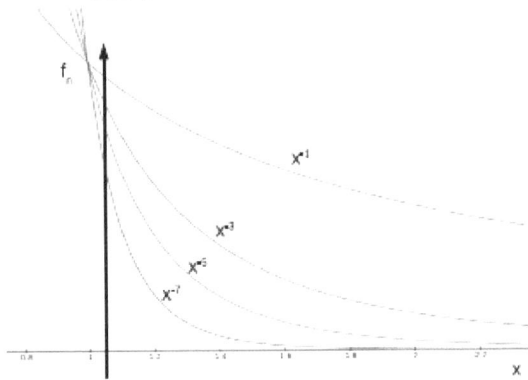

Fig. (8.2). Graph of the function $f_n(x) = \frac{1}{x^n}$.

8.2.1. Periodic Points

A **fixed point** or **invariant point** of a function f is an element $x \in D_f$ that is mapped to itself by the function *i.e.* $f(c) = c$.

Example 8.3. What is the fixed point of $f(x) = x^2$?

Solution 8.3. $c^2 = c \Leftarrow c^2 - c = 0 \Leftarrow c(c-1) = 0 \Leftarrow c = 0$ or $c = 1$. The points 0 and 1 are fixed points.

Definition 8.2. A point x is a **periodic point** of periodic p if: $f^p(x) = x$ and $\forall\, i \in \{1, 2, \cdots, p-1\}$: $f_i(x) \neq x$. ?, where $f^1 = f, f^2 = f \circ f, f^3 = f \circ f \circ f$ and so on.

Example 8.4. (i) Determine the **sequence of functions** of $f(x) = x^2$. (ii) Determine the **fixed points**.

Solution 8.4. (i) $\{f, f^1, f^2, \cdots, f^n\} = \{x^2, x^4, x^8, \cdots, x^n\}$. (ii) The **fixed points** are the solution of the equations $c^2 = c$, $c^4 = c$, $c^8 = c$, $c^n = c$. In all cases, the fixed points are 0 and 1.

8.3. POINTWISE AND UNIFORM CONVERGENCE

The convergence of function $f_k(x)$ can be **uniform** or **pointwise**. When it is a **uniform** convergence there is always a symmetry in the family of graphs of the function, unlike the **pointwise** convergence, as can be seen in Figs. (**8.3** and **8.4**).

8.3.1. Pointwise Convergence

Definition 8.3. A sequence of function $f_k(x)$ defined on D **converges pointwise** [88] on D if

$$\lim_{k \to \infty} f_k(x) \text{ exists for each point } x \text{ in } D.$$

Remark 8.2. See the conditions for the existence of a limit (Def. 3.1).

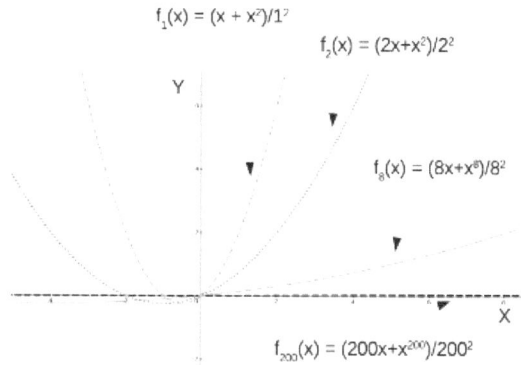

Fig. (8.3). Graph of **pointwise** convergence of the function $f_k(x) = \dfrac{kx + x^2}{k^2}$.

Note 8.1. Fig. (**8.3**) shows that the graphs of the functions **do not** converge to the x −axis at the same pace, therefore, they **are not** symmetric.

Example 8.5. Does the sequence of function $f_k(x) = nx$ converge pointwise? [88].

Solutuion 8.5. No, it does not because $\lim_{k \to \infty} kx = \infty$.

Example 8.6. Does the sequence of function $f_k(x) = \frac{x}{n}$ converge pointwise? ?.Yes, the sequence of the function converges to the constant function 0. For $\varepsilon > 0$ we have $N > \frac{x}{\varepsilon}$, then $|f_k(x) - 0| = \frac{x}{k} < \frac{x}{N} < \varepsilon$ for $k > N$ [18].

8.3.2. Uniform Convergence

Definition 8.4. The sequence of functions $f_k(x)$ converges to $f(x)$ **uniformly** on D if, and only if, for every $\varepsilon > 0$ there is an integer N such that for every $k \geq N$ and for every x on D, *i.e.*

$$|f_k(x) - f(x)| < \varepsilon \Leftarrow f_k(x) \tilde{A} f(x) \text{ on } D.$$

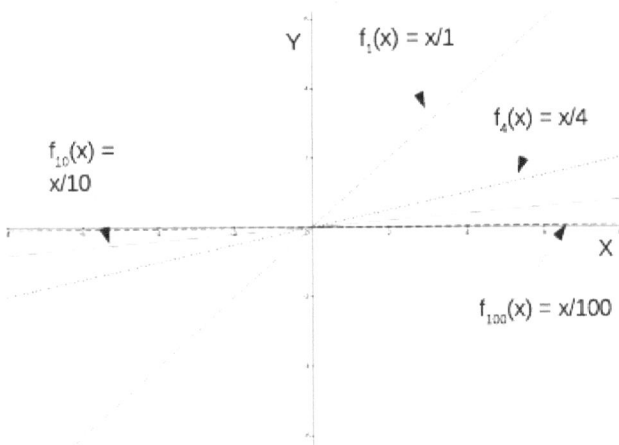

Fig. (8.4). Graph of **uniform** convergence of the function $f_k(x) = \frac{x}{k}$.

Note 8.2. Fig. (**8.4**) shows that the graphs of the functions **converge** to the x −axis at the same pace, therefore, these graphs are **symmetric**.

Example 8.7. Does the sequence of function $f_k(x) = e^{-nx}$ on [0,3] converge uniformly? [89].

Solution 8.7. Yes because it converges pointwise to constant function 0, then
$|f_k(x) - f(x)| = |f_k(x)| = \sup_{x \in [0,\infty]} |e^{-nx}| = \sup_{x \in [0,\infty]} e^{-nx} = e^{-n} \to$
0as$n \to \infty$.

8.3.3. Monotonicity

Definition 8.5. A **sequence of functions** $f_k(x)$ is **increasing** if $f_k(x) \le f_{k+1}(x)$ for all x and all k; it is **decreasing** if $f_k(x) \ge f_{k+1}(x)$ for all x and all k; and it is **monotone** if it is increasing or decreasing.

Example 8.8. Provide an example of (i) increasing sequence of functions, (ii) decreasing sequence of functions, and (iii) monotone sequence of functions.

Solution 8.8. (i) $f_k(x) = kx^2$. (ii) $f_k(x) = -kx^2$. (iii) $f_k(x) = (-1)^n x$.

8.3.4. Differentiation

If (f_k) is a sequence of differentiable functions on $[a, b]$, such that $\lim_{k \to \infty} f_k(x_0)$ exists (and is finite) for some $x_0 \in [a, b]$ and the sequence (f'_k) converges uniformly on $[a, b]$, then f_k converges uniformly to a function f on $[a, b]$ and $f'(x) = \lim_{k \to \infty} f'_k(x)$ for $x \in [a, b]$ (Definition taken from [90]).

8.4. PROPERTIES OF CONVERGENCE

Let the **convergent sequences of functions** $f_k \tilde{A} f \in \mathbb{R}$ and $g_k \tilde{A} g \in \mathbb{R}$, then [86]:

Property 1. $\alpha f_k = \alpha f$ (constant).
Property 2. $f_k + g_k = f + g$ (addition).
Property 3. $f_k - g_k = f - g$ (substraction).
Property 4. $(f_k)(g_k) = fg$ (multiplication).
Property 5. If $g \ne 0$, $(f_k)/(g_k) = \frac{f}{g}$ (division).
Property 6. $f_k(g) \tilde{A} f(g)$ (composition).

Remark 8.3. It is important to note that the preservation of the characteristics of **sequence of function** and the function only takes place with the **convergence**.

Theorem 8.1. *If the* **sequence of functions** f_k **converges** *to a function* f, *the following characteristics are preserved in both functions: odd, even, periodic, decreasing, increasing, and constant [86].*

8.5. SERIES OF FUNCTIONS

Definition 8.7. A **series of function** is a **series** of a **sequence of functions** $f_k(x)$ such that

$$\sum_{k=1}^{\infty} f_k(x).$$

Example 8.9. Provide an example of: (i) **Series of functions**. (ii) **Sequence of functions**. (iii) In both cases show their elements.

Solution 8.9. (i) $\sum_{k=0}^{\infty} (\frac{x}{1+x})^k$. (ii) $f_k(x) = \{kxe^{-kx}\}$. (iii) From $\sum_{k=0}^{\infty} (\frac{x}{1+x})^k$, its **sequence of partial sums** is $\{(\frac{x}{1+x})^0, (\frac{x}{1+x})^0 + (\frac{x}{1+x})^1, (\frac{x}{1+x})^0 + (\frac{x}{1+x})^1 + (\frac{x}{1+x})^2, \cdots\} = \{1, 1 + \frac{x}{1+x}, 1 + \frac{x}{1+x} + \frac{x^2}{(1+x)^2}, \cdots\}$; from $(\frac{x}{1+x})^k$, its elements are $\{0, xe^{-x}, 2xe^{-2x}, 3xe^{-3x}, \cdots\}$.

8.6. CASE STUDY: $\frac{kx}{1+k^2x^2}$

Case 8.1. Let the sequence of functions $f_k(x) = \frac{(kx)}{1+k^2x^2}$ on $[0, \infty)$ (Case taken with altered formatting [81]).

Since (f_k) converges pointwise to the constant function $f(x) = 0$ and $(1 + k^2x^2) \approx (k^2x^2)$, then $k \to \infty$.

$$\lim_{k \to \infty} f_k(x) = \lim_{k \to \infty} \frac{kx}{k^2x^2} = \frac{1}{x} \lim_{k \to \infty} \frac{1}{k} = 0$$

If $< \frac{1}{2}$, then replacing $x\frac{1}{k}$ in both functions

$$|f_k(\frac{1}{k}) - f(\frac{1}{k})| = \frac{1}{2} - 0 > \varepsilon.$$

From this result the sequence of functions $f_n(x) = \frac{(kx)}{1+k^2x^2}$ converges uniformly to the constant function $f(x) = 0$.

8.7. CASE STUDY: $\frac{\sin(kx+3)}{\sqrt{kx+1}}$

Case 8.2. Does the sequence of functions $f_k(x) = \frac{\sin(kx+3)}{\sqrt{kx+1}}, \forall x \in \mathbb{R}$ converge pointwise? (Case taken with altered formatting [81]).

$$\frac{-1}{\sqrt{k+1}} \leq \frac{\sin(kx+3)}{\sqrt{kx+1}} \leq \frac{-1}{\sqrt{k+1}}, \forall x \in \mathbb{R}$$

Since

$$\lim_{k \to \infty} \frac{-1}{\sqrt{k+1}} = 0$$

then, $\lim_{k \to \infty} f_k(x) = 0$.

The sequence of functions f_k, converges pointwise to the function $f(x) = 0$.

8.8. CASE STUDY: $\frac{kx^2+1}{kx+1}$

Case 8.3. In the graph of the **sequence of functions** $(\frac{kx^2+1}{kx+1})$ (Fig. **8.5**) [19], the uniform convergence **cannot** be observed in all the domain, however, the convergence **can** be observed in the interval $[1,2]$.

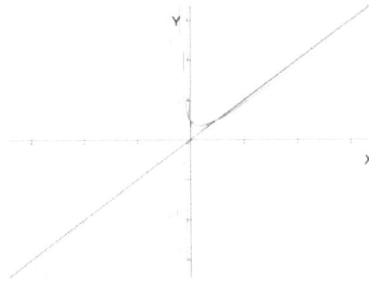

Fig. (8.5). Graph of $(f_k) = \frac{kx^2+1}{kx+1}$.

Expanding the interval we have (Fig. **8.6**)

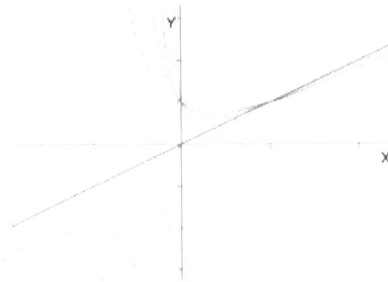

Fig. (8.6). Graph of $(f_k) = \frac{kx^2+1}{kx+1}$.

Analitically the convergence can be verified and we can say that f_k converges pointwise to $x \in [1,2]$.

$$\lim_{k\to\infty} \frac{kx^2 + 1}{kx + 1} = \lim_{k\to\infty} \frac{k^2 + \frac{1}{k}}{x + \frac{1}{k}} = x$$

To show the uniform convergence we verify the closeness to $|f_k - f|$,

$$|\frac{kx^2 + 1}{kx + 1} - x| = |\frac{1 - x}{kx + 1}| \le \frac{1 + |x|}{kx + 1} \le \frac{3}{1 + k}$$

Since $\frac{3}{1+k}$ tends to 0 as k tends to infinity, f_k converges uniformly to $x \in [1,2]$.

8.9. EXERCISES

Exercise 8.1. Define the sequence of function $f_k(x) = k$.

Exercise 8.2. List the elements of the sequence of functions in both intervals [91] and calculate $\lim_{k \to \infty} f_k(x)$.

$$f_k(x) = \begin{cases} x^k & for \quad x \in [0,1] \\ 1 & for \quad x \in (1,\infty) \end{cases} \tag{8.1}$$

Exercise 8.3. Does the sequence of functions f_k ? converge pointwise to the function f?

$$f_k(x) = \begin{cases} 1 - kx & for \quad 0 \le x \le \frac{1}{k} \\ 0 & for \quad \frac{1}{k} \le x \le 1 \end{cases} \tag{8.2}$$

$$f(x) = \begin{cases} 0 & for \quad x \in [0,1] \\ 1 & for \quad x = 0 \end{cases} \tag{8.3}$$

Exercise 8.4. Graph the sequence of functions (Ex. 8.2).

$$f_k(x) = \begin{cases} x^k & for \quad x \in [0,1] \\ 1 & for \quad x \in (1,\infty) \end{cases} \tag{8.4}$$

Exercise 8.5. Is the pointwise convergence of a sequence of functions related to the derivative of the sequence of functions? Give an example [93].

Exercise 8.6. What is the relationship between the uniform convergence and the pointwise convergence in a sequence of functions [93].

Exercise 8.7. Graph of the sequence of functions $f_n(x) = (xe^{-x})^n$.

Exercise 8.8. Are there non differentiable continuous functions at any point? [92]

Exercise 8.9. Show the first three terms of the series of functions $\sum_{n=1}^{n} (xe^{-}x)^n$ at the point $x = x_0$.

Exercise 8.10. Show the first term of the series of functions $(x_j e^{-x_j})_{j=1,3,5}^{n=1}$ at the points $x_j = \{1,3,5\}$.

CHAPTER 9

Antidifferentiation

Abstarct: This chapter studies the antiderivative function and its relation with the integral operator. The Fundamental Theorem of Calculus is reviewed and different integration techniques are exemplified. The series of functions named Fourier series and their application to geometrically approximate a function is also studied here.

Keywords: Antiderivative, Average value of a function, Fourier series Fundamental Theorem of Calculus, Improper integration, Integration by parts, Integration by substitution, Integration using Maps, Movement on a Line, Partial fraction decomposition, Techniques of antidifferentiation, Trigonometric integrals.

9.1. ANTIDERIVATIVE

Definition 9.1. The function $F(x) + c$ such that $F'(x) = f(x)$ $\forall x \in \mathbb{R}$, where c is a constant is named **antiderivative** continuous function (or indefinite integral) [94] and holds (Eq. 9.1).

$$F(x) - c = \int_0^x f(t) \ dt \qquad (9.1)$$

Note that there are many solutions for the function $F(x)$ that can be obtained by varying the constant $c \in \mathbb{R}$, all of them are translations of $F(x)$.

Remark 9.1. The procedure to determine the **antiderivative** function is named **antidifferentiation** and the operator $\int_0^x f(t) \ dt$ is known as the **indefinite integral**.

Example 9.1. Determine the antiderivative function $F(x)$ of the function $f(x) = x$.

Solution 9.1. The antiderivative $F(x)$ can be determined considering $F'(x) = x$, thus the function F is of higher degree, *e.g.* x^2, then it is divided by 2 and the constant is added, *i.e.* $F(x) = \frac{x^2}{2} + c$ and its derivative $F'(x) = (\frac{x^2}{2} + c)' = 2\frac{x}{2} + 0 = x = f(x)$.

It is not practical determining the antiderivative from scratch. There are equations to obtain it (Sect. 9.5). Here is the formula to find the antiderivative of a polynomial function (Eq. 9.2), so the reader can understand the following concepts.

Carlos Polanco
All rights reserved-© 2020 Bentham Science Publishers

$$F(x) = \int \alpha x^n = \alpha \frac{x^{n+1}}{n+1} + c, \text{ for } \alpha \in \mathbb{R} \text{ and } n \in \mathbb{N}. \tag{9.2}$$

Example 9.2. Find the antiderivative function $F(x)$ of the function $f(x) = 3x^4 + 2x^5 - 3$, using (Eq. 9.1) and (Eq. 9.2).

Solution 9.2. $F(x) = \int_0^x f(t) \, dt + c = \int_0^x 3t^4 + 2t^5 - 3 \, dt + c = [\frac{3}{5}t^5 + \frac{2}{6}t^6 - 3t]_0^x + c = \frac{3}{5}x^5 + \frac{1}{3}x^6 - 3x + c.$

Given a function $f(x)$, the procedure to determine its **derivative** is named **differentiation** and the process to determine its **antiderivative** is known as **antidifferentiation**, both procedures are inverse [95] (Eq. 9.3).

$$(\int f(x) \, dx)'_x = (F(x) + c)'_x = f(x). \tag{9.3}$$

Remark 9.2. Eq. 9.3 is the result of the **First Fundamental Theorem of Calculus** (Sect. 2), whose reading is recommended as it is the basis for Differential and Integral Calculus.

Particularly, if an **indefinite integral** is in a closed interval $[a, b] \in D_f$, it is named **definite integral** and it is equivalent to the **area** under the graph of the function (Eq. 9.4) in that closed interval (Fig. **9.1**).

This equivalence is true when the graph of the function is positive in the integration interval, otherwise, the integral will have a negative result, therefore, it is necessary to apply the absolute value to the integral in that interval.

$$Area = \int_a^b f(x) \, dx = F(b) - F(a) \tag{9.4}$$

Remark 9.3. Eq. 9.4 is the result of the **Second Fundamental Theorem of Calculus** (Sect. 9.3), its reading is recommended as it is the basis of Differential and Integral Calculus.

The definite integral can also be obtained calculating the limit of the series (Eq. 9.5).

$$Area = \int_a^b f(x) \, dx = \lim_{n\to\infty} \sum_{i=1}^n f(x_i)(x_i - x_{i-1}) = \lim_{n\to\infty} \frac{(b-a)}{n} \sum_{i=1}^n f(a + i\frac{(b-a)}{n}) \tag{9.5}$$

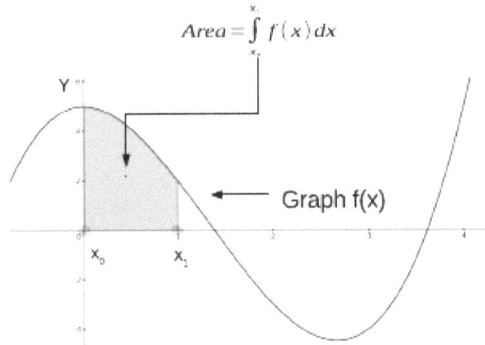

$$Area = \int_{x_o}^{x_2} f(x)\,dx$$

Graph f(x)

Fig. (9.1). Graph of the function $f(x) = x^3 - 4x^2 + 5$ and its closed area in $[a, b]$.

The limit of the series represents the sum of the areas of the n rectangles formed over the n subintervals in [a, b]. The subintervals are of equal length and the height of the ith rectangle is the value of the function $f(x)$ at a chosen sample point in the ith subinterval. The width of each rectangle is $\frac{b-a}{n}$ and the height of the rectangle in the ith subinterval is given by $f(a + i\frac{(b-a)}{n})$, where $a + i\frac{(b-a)}{n}$ is a sample point in the ith subinterval [94].

Example 9.3. Be the function $f(x) = x$ in the interval $[1,2]$. (i) Compute the closed area in $[1,2]$ using (Def. 7.2). (ii) Compute the closed area in the same interval using (Eq. 9.4).

Solution 9.3. (i)

$$
\begin{aligned}
Area &= \lim_{n\to\infty} \frac{(b-a)}{n} \sum_{i=1}^{n} f(a + i\frac{(b-a)}{n}) = \lim_{n\to\infty} \frac{(2-1)}{n} \sum_{i=1}^{n} f(1 + i\frac{(2-1)}{n}) \\
&= \lim_{n\to\infty} \frac{1}{n} \sum_{i=1}^{n} f(1 + \frac{i}{n}) = \lim_{n\to\infty} \frac{1}{n} \sum_{i=1}^{n} (1 + \frac{i}{n}) \qquad\qquad (9.6)\\
&= \lim_{n\to\infty} \frac{1}{n} [\sum_{i=1}^{n} 1 + \frac{1}{n}\sum_{i=1}^{n} i] = \lim_{n\to\infty} \frac{1}{n} [n + \frac{n^2+n}{2n}] = \lim_{n\to\infty} \frac{3n+1}{2n} = \frac{3}{2}.
\end{aligned}
$$

(ii) $\int_{1}^{2} f(x)\ dx = \int_{1}^{2} x\ dx = [\frac{x^2}{2}]_{1}^{2} = \frac{4}{2} - \frac{1}{2} = \frac{3}{2}.$

Note 9.1. For a detailed review of this series see (Case 7.11).

Example 9.4. Obtain the (i) antiderivative function $F(x)$ of the function $f(x) = x^3 - 4x^2 + 5$ using (Def. 9.2). (ii) From (Eq. 9.4) determine the closed area in

[0,1]. (i) $F(x) = \int x^3 - 4x^2 + 5 \ dx = \frac{1}{4}x^4 - \frac{4}{3}x^3 + 5x + c$. (ii) The closed area is the function $F(x)$ (without the constant), evaluated in the points $x = 0$ and $x = 1$, which is the area $= F(1) - F(0) = \frac{47}{12}$.

9.2. FIRST FUNDAMENTAL THEOREM OF CALCULUS

The Fundamental Theorem of Calculus states that the derivative of the antiderivative function $F'(x)$ of the continuous function $f(x)$ is the $f(x)$ itself, this makes possible to obtain the antiderivative function $F(x)$ with the indefinite integral operator. Both antiderivatives will differ only by the constant c.

Definition 9.2. Let the function $f(x)$ be continuous in the closed interval $[a, b]$ and defined $F(x) = \int_a^x f(t) \ dt, x \in [a, b]$; then $F'(x) = \frac{d}{dx}[\int_a^x f(t) \ dt] = f(x)$ (**Definition taken from** [94]).

Proof. Let function $f(x)$ be continuous in $[a, b]$ and $F(x) = \int_a^x f(t) \ dt, \forall x \in [a, b]$, then $\quad F'(x) = \lim_{\Delta x \to 0} \frac{f(x+\Delta x) - f(x)}{\Delta x} \Leftrightarrow \lim_{\Delta x \to 0} \frac{\int_a^{x+\Delta x}}{f}(t) \ dt - \int_a^x f(t) \ dt \Delta x = \frac{1}{\Delta x}\int_x^{x+\Delta x}$
$f(t)dt = \lim_{\Delta x \to 0} \frac{f(\alpha)}{x+\Delta x - x}$, because $x \leq \alpha \leq x + \Delta x + x \Rightarrow \alpha = x$. Since $F'(x) = \lim_{\Delta x \to 0} f(\alpha)$, then $\frac{d}{dx}F'(x) = f(x)$.

Example 9.5. If $f(x) = 3x^2$, verify (Thm. 9.2).

Solution 9.5. $F(x) = \int_0^x f(t) \ dt = \int_0^x 3t^2 \ dt = [t^3]_0^x = x^3 + c$, then $F'(x) = (x^3 + c)' = 3x^2$.

9.3. SECOND FUNDAMENTAL THEOREM OF CALCULUS

Let the function $f(x)$ be continuous in the closed interval $[a, b]$, then $\int_a^b f(x) \ dx = F(b) - F(a)$ such that $F'(x) = f(x)$, where $F(x)$ is any antiderivative of $f(x)$ in $[a, b]$ (**Definition taken from** [94]).

Proof. $F'(x) = f(x) \Leftrightarrow F(x) = \int_a^x f(t) \, dt + c$. Since $F(a) = \int_a^a f(t) \, dt$ is 0, then $F(a) = c \Rightarrow x = b$. So $F(b) = \int_a^b f(t) \, dt + F(a) \Rightarrow \int_a^b f(t) \, dt = F(b) - F(a)$.

9.4. NOTATION

Hereafter, we will name the **integral** of $f(x)$, $\int f(x) \, dx$ as the **integrand** and the symbol dx will be the **differential** of the variable x.

Example 9.6. If $F(x) = x^3 + x^2 + x$ is the antiderivative of $f(x)$ in the interval $[0,5]$. What is the area of (x) in that interval?

Solution 9.6. $F(x) = \int_a^b f(x) \, dx = F(b) - F(a) = \int_0^1 f(x) \, dx = F(1) - F(0) = 3$.

9.5. BASIC ANTIDERIVATIVES EQUATIONS

Since it is always possible to find the **derivative** of a continuous function, it **is not** the same for the **antiderivative** (or indefinite integral) of a function $f(x)$. Here you will find some basic antiderivative equations that will be used in this section.

Rule 1. $F(x) = \int \alpha \, dx = \alpha + c$, for $\alpha \in \mathbb{R}$.
Rule 2. $F(x) = \int \alpha x^n \, dx = \alpha \frac{x^{n+1}}{n+1} + c$, for $\alpha \in \mathbb{R}$ and $n \in \mathbb{N}$.
Rule 3. $F(x) = \int \sin x \, dx = -\cos x + c$.
Rule 4. $F(x) = \int \cos x \, dx = \sin x + c$.
Rule 5. $F(x) = \int e^x \, dx = e^x + c$.
Rule 6. $F(x) = \int \frac{1}{x} \, dx = \ln|x| + c$.
Rule 7. $F(x) = \int \frac{1}{\sqrt{1-x^2}} \, dx = \sin^{-1}x + c$.

Example 9.7. Provide examples of definite integrals using the Rules 5.

Solution 9.7. (i) $\int_1^3 5 \, dx = 5 \int_1^3 dx = 5[x]_1^3 = 10$. (ii) $\int_1^3 5x \, dx = 5[\frac{x^2}{2}]_1^3 = 20$. (iii) $\int_{-\pi}^{\pi} \sin x \, dx = [-\cos x]_{-\pi}^{\pi} \, dx = -\cos\pi + \cos(-\pi) = 0$ (Nt. 11). (iv) $\int_{-\pi}^{\pi} \cos x \, dx = [\sin x]_{-\pi}^{\pi} \, dx = \sin\pi - \sin(-\pi) = 2\sin\pi = 0$ (Nt.   11).   (v) $\int_0^1 e^x \, dx = e^x]_0^1 = e - 1$. (vi) $\int_1^2 \frac{1}{x} \, dx = [|\ln x|]_1^2 = \ln 2$. (vii) $\int_0^{\pi} \frac{1}{\sqrt{1-x^2}} \, dx = [\sin^{-1}(x)]_0^{\pi} = \sin^{-1}(\pi)$.

9.6. PROPERTIES OF ANTIDERIVATIVES

Let the continuous functions $f_1 : \mathbb{R} \to \mathbb{R}$ and $f_2 : \mathbb{R} \to \mathbb{R}$, if $\int_a^b f_1(x)\,dx \in \mathbb{R}$, $\int_a^b f_2(x)\,dx \in \mathbb{R}$ and $\alpha \in \mathbb{R}$ then:

Property 1. $\int_a^b f(x)\,dx = -\int_b^a f(x)\,dx$ (reversing).
Property 2. $\int_a^b f_1(x) + f_2(x)\,dx = \int_a^b f_1(x)\,dx + \int_a^b f_2(x)\,dx$ (addition).
Property 3. $\int_a^b f_1(x) - f_2(x)\,dx = \int_a^b f_1(x)\,dx - \int_a^b f_2(x)\,dx$ (substraction).

Example 9.8. Provide examples of definite integrals using the Rules 9.6.

Solution   9.8.   (i) $\int_0^1 x^2 + 3x^5\,dx = [\frac{x^3}{3} + \frac{x^6}{2}]_0^1 = 1 + \frac{1}{2} = \frac{3}{2} = -\int_1^0 x^2 + 3x^5\,dx = \frac{x^3}{3}$
$+ \frac{x^6}{2}]_1^0 = -1 - \frac{1}{2}$. (ii) $\int_0^\pi \sin x + \cos x\,dx = [-\cos x + \sin x]_0^\pi = 1 = \int_0^\pi \sin x\,dx + \int_0^\pi \cos x\,dx$.
(iii) $\int_2^3 e^x\,dx - \int_2^3 \sqrt{x}\,dx = \int_2^3 e^x - \sqrt{x}\,dx = [e^x - \frac{1}{2\sqrt{x}}]_2^3$.

9.7. MAIN TECHNIQUES OF ANTIDIFFERENTIATION

In this section, only the most common techniques to solve an integral will be presented.

9.7.1. Integration by Substitution

This procedure consists of expressing the definite integral in terms of the variable u, changing the variable (Eq. 9.7).

$$\int_a^b f(x)\,dx = \int_{g(a)}^{g(b)} f(g(u))\,du \qquad (9.7)$$

Remark 9.4. Note that this procedure requires the substitution of all elements in the variable x for the variable u.

Example 9.9. Compute the definite integral $\int_0^2 2x\cos x^2\,dx$.

Solution   9.9.   If $u = x^2$, $du = 2xdx$ and $u_1 = (0)^2 = 0$, $u_2 = (2)^2 = 4$.
$\int_0^2 2x\cos x^2\,dx = \int_0^4 \cos u\,du = [\sin u\ ]_0^4 = \sin 4$.

9.7.2. Integration by Parts

This procedure is suggested when the integrand is a product of different types of functions, *e.g.* polynomial and exponential (Eq. 9.8).

$$\int_a^b u(x)v'(x)\ dx = [u(x)v(x)]_a^b - \int_a^b u'(x)v(x)\ dx \qquad (9.8)$$

Example 9.10. Compute the definite integral $\int_1^2 x\ln x\ dx$.

Solution 9.10. If $u(x) = \ln x$, $dv = x\ dx$, then $u'(x) = \frac{1}{x}\ dx$ and $v(x) = \int x\ dx = \frac{x^2}{2} + c$. $\int_1^2 x\ln x\ dx = [\frac{x^2}{2}\ln x]_1^2 - \int_1^2 (\frac{x^2}{2})(\frac{1}{x})\ dx = [\frac{x^2}{2}\ln x]_1^2 - [\frac{x^2}{4}]_1^2$.

9.7.3. Trigonometric Integrals

The trigonometric integrals (Eq. 9.9), (Eq. 9.10) can be used in a recursive way to obtain any integral in the integrand sines and cosines.

$$\int_a^b \sin^n x\ dx = -\frac{\sin^{n-1}x\cos x}{n} + \frac{n-1}{n}\int_a^b \sin^{n-2}x\ dx \qquad (9.9)$$

$$\int_a^b \cos^n x\ dx = \frac{\cos^{n-1}x\sin x}{n} + \frac{n-1}{n}\int_a^b \cos^{n-2}x\ dx \qquad (9.10)$$

Particularly (Eq. 9.11) and (Eq. 9.12).

$$\int_a^b \sin x\ dx = -\cos x \qquad (9.11)$$

$$\int_a^b \cos x\ dx = \sin x \qquad (9.12)$$

Remark 9.5. The (Eq. 9.11) and (Eq. 9.12) come from the (Eq. 9.9), (Eq. 9.10).

Example 9.11. Solve the (Eq. 9.9) for n = 2.

Solution 9.11. $\int_a^b \sin^2 x\ dx = [-\frac{\sin x\cos x}{2}]_a^b + \frac{1}{2}(b-a)$. From (Nt. 7.3) $\int_a^b \sin^2 x\ dx = [-\sin 2x]_a^b + \frac{1}{2}(b-a)$.

Note 9.2. $\sin 2x = 2\sin x \cos x$.

9.7.4. Partial Fraction Decomposition

This technique consists of breaking the expressions of the form $(ax + b)^k$ into series.

$$(ax + b)^k = \frac{A_1}{ax+b} + \frac{A_2}{(ax+b)^2} + \cdots \frac{A_k}{(ax+b)^k}, k \in \mathbb{N}. \tag{9.13}$$

Example 9.12. Show the integral $\int_a^b \frac{6x+13}{x^2+5x+6} \, dx$ in factors.

Solution 9.12. $\int_a^b \frac{6x+13}{x^2+5x+6} \, dx = \int_a^b \frac{1}{x+2} \, dx + \int_a^b \frac{5}{x+3} \, dx.$

9.7.5. Improper Integration

A definite integral is an **improper integral** if the integration interval is not finite, *i.e.* if the integration approaches to infinite, or if the function to integrate is non-continuous on the integration interval [96]. This particular definite integral $\int_a^\infty f(x) \, dx$ is equal to the limit of the integral $\lim_{\alpha \to 0} \int_a^\alpha f(x) \, dx$ (Eq. 9.14)

$$\int_a^\infty f(x) \, dx = \lim_{\alpha \to 0} \int_a^\alpha f(x) \, dx. \tag{9.14}$$

Example 9.13. Compute $\int_1^\infty \frac{1}{x^2} \, dx$

Solution 9.13. $\lim_{\alpha \to \infty} \int_1^\alpha \frac{1}{x^2} \, dx = \lim_{\alpha \to \infty} [-\frac{1}{x}]_1^\alpha = \lim_{\alpha \to \infty} 1 - \frac{1}{\alpha} = 1.$

9.8. INTEGRATION USING MAPS

This procedure consists of mapping $T: \mathbb{R} \to \mathbb{R}$ of the function $f(x)$ (Eq. 9.15)

$$\int_a^b f(x) \, dx \; = \int_{T(a)}^{T(b)} f(x) \circ T(u) \, |J(T(u))| \, du \tag{9.15}$$

where the Jacobian determinant $J(T)$ (Eq. 9.16) is the determinant of the map $T(u)$.

$$J(T(u)) = \left|\frac{dT}{du}\right| \tag{9.16}$$

Example 9.14. Consider the map $T: \mathbb{R} \rightarrow \mathbb{R}, 3u$. What is the Jacobian determinant?

Solution 9.14. $J(T(u)) = \left|\frac{dT}{du}\right| = 3$.

Example 9.15. Let the function $f(x) = x^2\sqrt{x^3}, x \in [0,2]$. (i) What is the area under the curve of that interval? (ii) What map is convenient to simplify this integral? (iii) Solve the integral using the map $T(u)$. (iv) Solve the integral using the change of variable method $u = x^3$. (v) Do the geometric analysis of both graphs.

Solution 9.15. (i) $\int_0^2 x^2\sqrt{x^3}\,dx = \left[\frac{2}{9}x^{\frac{9}{2}}\right]_0^2 = 5$. (ii) Let $T(u) = u^{\frac{1}{3}}, u \in [0,8]$. (iii)

If $\quad T(u) = u^{\frac{1}{3}}, \quad J(T(u)) = \frac{1}{3u^{\frac{2}{3}}}, \quad \int_0^2 x^2\sqrt{x^3}\,dx = \int_0^8 f \circ T(u)|J(T(u)|\,du = \frac{1}{3}$

$\int_0^8 u^{\frac{2}{3}}\sqrt{u}\ u^{-\frac{2}{3}}\,du = \frac{1}{3}\int_0^8 \sqrt{u}\,du = 5$. (iv) If $u = x^3 \Rightarrow du = 3x^2\,dx \Rightarrow dx = \frac{du}{3x^2}$

$\Rightarrow \int_0^8 x^2\sqrt{u}\ \frac{du}{3x^2} = \frac{1}{3}\int_0^8 \sqrt{u}\,du = 5$. (v) The closed areas of both functions $f_1(x) = x^2\sqrt{x^3}$ and $f_2(x) = \frac{1}{3}\sqrt{x}$, in different integration domains, are equivalent (Fig. **9.2**).

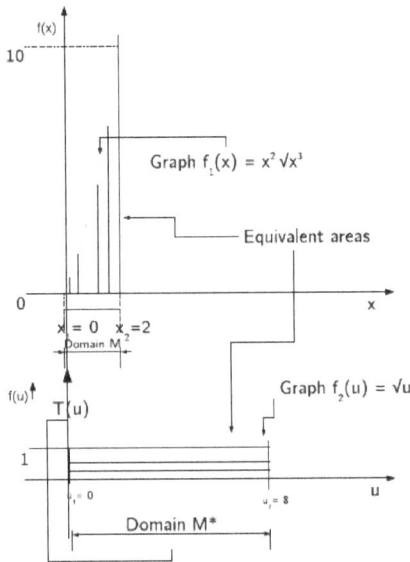

Fig. (9.2). The T mapping goes from M^* space to M space. The area under $f_1(x) = x^2\sqrt{x^3}$ over $[0,2]$ and the area under $f_2(u) = \sqrt{u}$ over $[0,8]$ are equivalent (Figure adapted from [29]).

9.9. CASE STUDY: AVERAGE VALUE OF A FUNCTION

*Case 9.1.*Let the function $f(x) = x^3 + 1$, determine the average value in the interval $[0,3]$.

The average value of the function $f(x)$ is given by $f_m(x) = \frac{1}{b-a}\int_a^b f(x)\ dx$, so when substituting (Eq. 9.17)

$$f_m(x) = \frac{1}{3-0}\int_0^3 x^3 + 1\ dx = \frac{1}{3}[\frac{1}{4}x^4 + 2x]_0^3 = 7.25. \tag{9.17}$$

What does $f_m(x) = 7.25$ represent?

If $f(x) = 7.25$, translating it into the original function $f(x) = x^3 + 1$, then $7.25 = x^3 + 1 \Rightarrow x = 1.84$

See (Fig. **9.3**), note that the rectangle $ABCD$ with 7.25 height and 3 width is equal to $7.25 \times 3 = 21.75$, which is coincident with the value of the area under the graph of the function $x^3 + 1$ in the interval $[0,3]$ (Eq. 9.18).

$$\int_0^3 x^3 + 1\ dx = [\frac{1}{4}x^4 + 2x]_0^3 = \frac{87}{4} = 21.75 \tag{9.18}$$

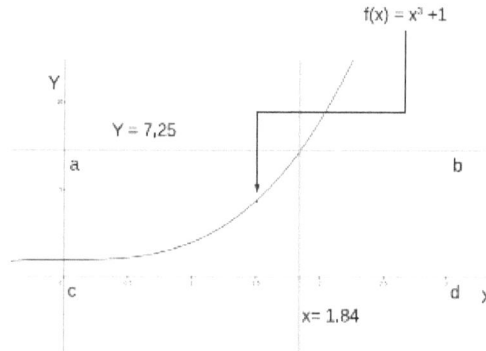

Fig. (9.3). Graph of the functions $f(x) = x^3 + 1$, $y = 7.25$, and $x = 1.84$ in the interval $[0,3]$.

9.10. CASE STUDY: DISPLACEMENT AND DISTANCE TRAVELED

Case 9.2. The displacement and the distance travelled seem synonymous but they are not. Suppose that, at some point, an object advances certain distance and then it returns the same distance. What is the displacement? The displacement is zero, because both distances were the same. However, the distance is not zero, therefore, both concepts are not equivalent.

Can we use the definite integral to obtain the same results?

Thus, $displacement = \int_{t_i}^{t_f} f(t) \ dt$, where t_i and t_f represent the time lapse and $f(t)$ is the function whose graph is the displacement.

Whereas $distance \ travelled = \int_{t_i}^{t_f} |f(t)| \ dt$, where t_i and t_f is the time lapse and $f(t)$ is the function whose graph represents the displacement.

Consider that the route is the graph of function $f(x) = 4 - t$ and that it took someone 4 secs to cover that distance forward and 4 seconds backward, then

$$displacement = \int_{t_i}^{t_f} f(t) \ dx = \int_0^8 4 - t \ dx = 4t - \frac{t^2}{2}]_0^8 = 0, \text{ whereas}$$

$$distancetravelled = \int_{t_i}^{t_f} |f(t)| \ dx = \int_0^8 |4 - t| \ dx = |4t| + |\frac{t^2}{2}|]_0^8 = 64.$$

9.11. CASE STUDY: FOURIER SERIES

Case 9.3. Let's use here the particular case of series of functions, because it uses definite integrals (Ch. 7).

Fourier series is a series of functions (Eq. 9.19) that enables us to approximate by superposition of sine and cosine functions, the function $f(x)$.

$$f(x) = \frac{a_0}{2} + \sum_{n=1}^{\infty} a_n \frac{\cos n\pi x}{L} + a_n \frac{\sin n\pi x}{L}, n \in \mathbb{N} \qquad \textbf{(9.19)}$$

where

$$a_n = \frac{1}{L} \int_{-L}^{L} f(x)\cos\frac{n\pi x}{L} \ dx$$

$$b_n = \frac{1}{L} \int_{-L}^{L} f(x)\sin\frac{n\pi x}{L} \ dx$$

Let's approximate the function $f(x) = x^2$ for $x \in [-\pi, \pi]$ with a Fourier series, where $a_0 = \frac{1}{\pi} \int_{-\pi}^{\pi} x^2 \cos\frac{\pi x}{\pi} \, dx = \frac{1}{\pi} [\frac{x^3}{3}]_{-\pi}^{\pi} = \frac{2}{3}\pi^2.$ $a_n = \frac{1}{\pi} \int_{-\pi}^{\pi} x^2 \cos nx \, dx = \frac{1}{\pi} \int_{pi}^{\pi} x^2 \cos nx \, dx$ (Eq. 9.8 with $u = x^2$ and $dv = \cos nx \, dx$), we obtain $\int_{pi}^{\pi} x^2 \cos nx \, dx = -\frac{4}{n\pi} x \sin nx \, dx.$ We apply again (Eq. 9.8 with $u = x$ and $dv = \sin nx \, dx$) in the last integral, then $a_n = \frac{4}{n^2\pi} [\pi \cos n\pi - \frac{\sin nx}{n}].$ Since the function $f(x)$ is even (Nt. 9.3), $b_n = 0$ and (Nt. 11), $a_n = \frac{4}{n^2}(-1)^n.$

Note 9.3. f is **even** if, and only if, $f(-x) = -f(x)$ and f is **odd** if, and only if, $f(-x) = f(x)$.

Note 9.4. $\sin n\pi = 0$ and $\cos n\pi = (-1)^n$ for integer $n \in \mathbb{N}$.

Now, let's substitute the three parameters (Eq. 9.19) considering (Nt. 9.4)

$$f(x) = \frac{\pi^2}{3} + \sum_{n=1}^{\infty} \frac{4}{n^2}(-1)^n \cos nx, n \in \mathbb{N}$$

Observe (Fig. **9.4**) where there are two approximations to the function $f(x) = x^2$.

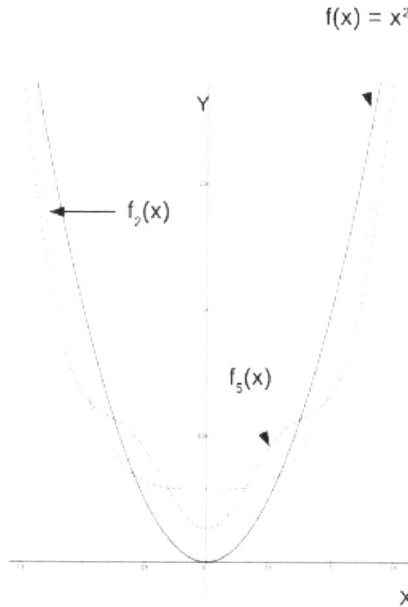

f(x) = x²

Fig. (**9.4**). Graph of function $f(x) = x^2$ and Fourier approximation for $n = 2$, and 5.

9.12. EXERCISES

Exercise 9.1. Compute the antiderivative $F(x)$ of the function $f(x) = \sin 2x$.

Exercise 9.2. Compute the area of function $f(x) = x$ in the interval $[-1,1]$.

Exercise 9.3. Can we determine the integral $\int_0^\pi f(x) \ dx$, without function $f(x)$, but knowing that $F(0) = a$ and $F(\pi) = b$?

Exercise 9.4. Compute $\int_1^2 xe^{x^2} \ dx$.

Exercise 9.5. Compute the indefinite integral $\int x^2 \cos x \ dx$.

Exercise 9.6. Compute the definite integral $\int_0^1 \frac{1}{2+x} \ dx$.

Exercise 9.7. We know that $\int_1^2 x^3 \ dx = \frac{1}{2}\int_1^4 u \ du$. Explain the reason for this equality.

Exercise 9.8. Compute the improper integral $\int_0^1 \frac{1}{\sqrt{x}} \ dx$.

Exercise 9.9. Compute the Jacobian determinant of the map $T(u) = u^3 e^u$.

Exercise 9.10. Let the function $f(x) = \sqrt{x^3} \ x \in [0,2]$. (i) What is the area under the curve of that interval? (ii) What map is convenient to simplify this integral? (iii) Solve the integral using the map $T(u)$.

CHAPTER 10

SOLUTIONS

Chapter 1

Solution 1.1. See (Fig. **1.9**)

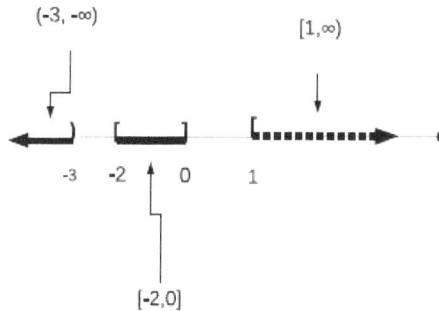

Fig. (1.9). Representation of intervals $(-\infty, -3)$, $[-2,0]$, and $[1, \infty)$ on the real oriented line.

Solution 1.2.

$$
\begin{aligned}
-a + (a + b) &= -a + (a + c) \\
(-a + a) + b &= (-a + a) + c \\
0 + b &= 0 + c \\
b &= c
\end{aligned}
\tag{1.1}
$$

Solution 1.3. Yes it is, because $\exists\, k = -2 \in \mathbb{Z} \supset$ -- $mk = (-2)(-2) = 4 = n$. In case that $k \notin \mathbb{Z}$, we say that m does not divide n.

Solution 1.4. (i) $\frac{x-3}{x+2} = 2 \Leftrightarrow x + 3 = 2(x + 2) \Leftrightarrow x = -7$. (ii) $\frac{x-3}{x+2} = -2 \Leftrightarrow x + 3 = -2(x + 2) \Leftrightarrow x = -\frac{1}{3}$. The solution is $\{-7, -\frac{1}{3}\}$.

Solution 1.5. $\sqrt{2x + 2} + x = \Leftarrow \sqrt{2x + 2} = 1 - x \Leftarrow x^2 - 4x - 1 = 0$. Solutions $x_1 = 2 + \sqrt{5}$, $x_2 = 2 - \sqrt{5}$.

Carlos Polanco
All rights reserved-© 2020 Bentham Science Publishers

Solution 1.6.

$$
\begin{aligned}
10x + 5 &\leq 25 \\
10x &\leq 20 \\
x &\leq \frac{20}{10} \\
x &\leq 2.
\end{aligned}
\tag{1.2}
$$

Solution 1.7.

$$
\begin{aligned}
7 &\leq 3x - 2 \leq 13 \\
9 &\leq 3x \leq 15 \\
3 &\leq x \leq 5.
\end{aligned}
\tag{1.3}
$$

Solution 1.8.

$$
\begin{aligned}
\frac{3}{x} &< 5 \,(\text{if } x > 0) \\
3 &< 5x \\
\frac{3}{5} &< x.
\end{aligned}
\tag{1.4}
$$

$$
\begin{aligned}
\frac{3}{x} &< 5 \,(\text{if } x < 0) \\
3 &> 5x \\
\frac{3}{5} &> x.
\end{aligned}
\tag{1.5}
$$

From (Eqs. 1.4 and 1.5), the solutions are $x > \frac{3}{5}$ and $x < 0$, then $(-\infty, 0) \cup (\frac{3}{5}, +\infty)$.

Solution 1.9.

$$
\begin{aligned}
\frac{2x+1}{3x-6} &\geq 3 \,(\text{if } 3x - 6 > 0 \Rightarrow x > 2) \\
2x + 1 &\geq 3(3x - 6) \\
2x + 1 &\geq 9x - 18 \\
-7x &\geq -19 \\
7x &\leq 19 \\
x &\leq \frac{19}{7}.
\end{aligned}
\tag{1.6}
$$

$$\begin{aligned}
\frac{2x+1}{3x-6} &\geq 3 \text{(if } 3x - 6 < 0 \Rightarrow x < 2) \\
2x + 1 &\leq 3(3x - 6) \\
2x + 1 &\leq 9x - 18 \\
-7x &\leq -19 \\
7x &\geq 19 \\
x &\geq \frac{19}{7}.
\end{aligned}$$

(1.7)

From (Eqs. 1.6 and 1.7), the solutions are (i) $x > 2$ and $x \leq \frac{19}{7}$, (ii) $x < 2$ and $x \geq \frac{19}{7}$, therefore, there is no element in common.

Solution 1.10. $x^2 + x - 6 > 0 \Leftarrow (x + 3)(x(x - 2) > 0$. (i) Since both terms are positive $x > -3$ and $x > 2$, the solution is $x > 2$. (ii) Since both terms are negative $x < -3$ and $x < 2$, the solution is $x < -3$. For (i) and (ii) the general solution is $x \in (-\infty, -3) \cup (2, \infty)$.

Solution 1.11. The **lower bounds** in the interval is $[-\infty, -6)$, the **upper bounds** is the interval is $[6, \infty)$, the **infimum** element is -6, and the **supremum** element is 6.

Solution 1.12. The set has no **lower bounds**, the **upper bounds** is the interval $[8, \infty)$, there is no **infimum** element, and the **supremum** element is 8.

Solution 1.13. The **lower bounds** in the interval is $[-\infty, -3)$, the **upper bounds** is the interval $(4, \infty)$, the **infimum** element is -3, and the **supremum** element is 4.

Solution 1.14. From (i) $|x - 2| < 1 \Leftarrow -1 < x - 2 < 1$. From (ii) $|x^2 - 4| < 5 \Leftarrow -5 < x^2 - 4 < 5 \Leftarrow -1 < x^2 < 9 \Leftarrow 1 < x < 3$. (i) and (ii) are equivalent.

Solution 1.15. $|x - 1| < |x| \Leftarrow (x - 1)^2 < x^2 \Leftarrow x^2 - 2x + 1 < x^2 \Leftarrow 2x > 1 \Leftarrow x > \frac{1}{2}$. The solution is $x \in (\frac{1}{2}, \infty)$.

Solution 1.16. (i) If $x = 0 \Rightarrow |0| = |-0| \Leftarrow 0 = 0$. (ii) if $x > 0 \Rightarrow x = x$. (iii) If $x < 0$, then $-x > 0 \Rightarrow -x = -x$. So $|x| = |-x|, \forall x \in \mathbb{R}$.

Solution 1.17. $-\frac{3}{5} < 5x - 1 < \frac{3}{5} \Leftarrow -\frac{8}{5} < 5x < \frac{8}{5} \Leftarrow -\frac{8}{25} < 5x < \frac{8}{25}$.

Chapter 2

Solution 2.1. (i) f(oldperson) = 80years, f(youngperson) = 20years. (ii) Yes it is. f is a function because its rule is well defined.

Solution 2.2. (i) $(f+g)(x) = f(x) + g(x) = \sqrt[3]{x+3} + \frac{1}{x} = \frac{x\sqrt[3]{x+3}+1}{x}$. $(f - g)(x) = f(x) - g(x) = \sqrt[3]{x+3} - \frac{1}{x} = \frac{x\sqrt[3]{x+3}-1}{x}$. (iii) $(fg)(x) = f(x) \cdot g(x) = \frac{\sqrt[3]{x+3}}{x}$.

Solution 2.3. (i) $f: \mathbb{R} \to \mathbb{R}^+, \sqrt[5]{|x|}$. (ii) The domain is $\mathbb{R}$. (iii) The image is $\mathbb{R}^+$. (iv) Its graph is (Fig. **2.22**).

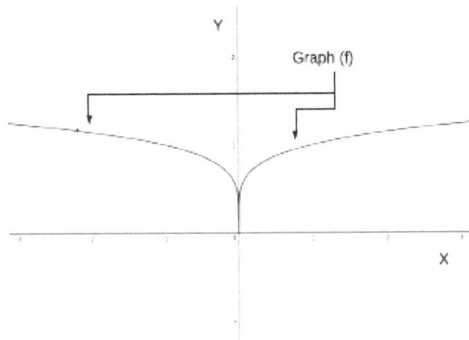

Fig. (2.22). The graph of function $\sqrt[5]{|x|}$. Graph plotter [36].

Solution 2.4. (i) Yes, it is because $\forall\ x_1, x_2 \in (-\infty, \infty), f(x_1) \neq f(x_2) \Rightarrow x_1 \neq x_2$. (ii) Its graph is (Fig. **2.23**).

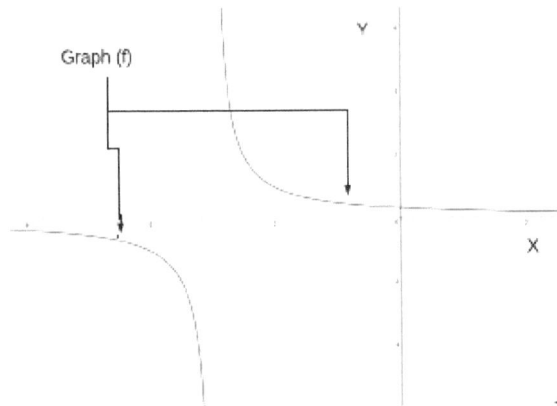

Fig. (2.23). The graph of function $\frac{1}{x+3}$. Graph plotter [36].

Solution 2.5. No, it is not. This is one of the most common mistakes. For instance, let the function $f(x) = x$ then $f^{-1}(y) = y. \frac{1}{y}$ is not correct.

Solution 2.6. (i) $f \circ g = f(g(x)) = f(\ln x) = \ln x - 3$. (ii) $g \circ f = g(f(x)) = g(x - 3) = \ln x - 3$.

Solution 2.7. (i) $f^{-1}(x) = y + 2$. (ii) $f \circ f^{-1} = f(f^{-1}) = f(y + 2) = y + 2 - 2 = y$. (iii) $f^{-1} \circ f = f^{-1}(f) = f^{-1}(x - 2) = x - 2 + 2 = x$. (iv) If $f \circ f^{-1} = f^{-1} \circ f$, then $f^{-1}(x)$ is the inverse function of f.

Solution 2.8. (i) $T_2 \circ T_1(x) = T_2(\frac{x}{2}) = (\frac{x}{2}, \frac{x}{2})$, $t \in [0,2]$. (ii) Its graph is (Fig. **2.24**).

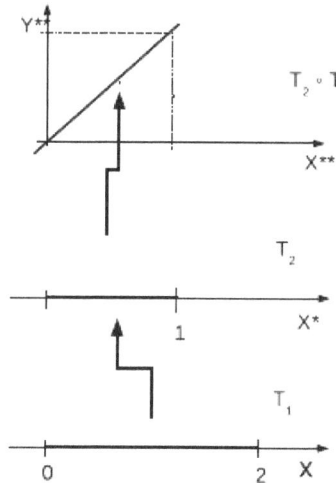

Fig. (2.24). Description of the composition $T_2 \circ T_1$. Graph plotter [36].

Solution 2.9. (i) $f_2 \circ f_1(x) = f_2(f_1) = f_2(\sin 2x) = \cos(\sin 2x)^2 = \cos(\sin^2 2x)$. (ii) $f_1 \circ f_2(x) = f_1(f_2) = f_1(\cos x^2) = \sin(\cos x^2)$.

Solution 2.10. (i) No, it is not because to each element of $D_f = [-1,1]$ corresponds two elements of I_f. (ii) $T: [0,2\pi] \subset \mathbb{R} \rightarrow \mathbb{R}^2$, $(\cos t, \sin t)$.

Chapter 3

Solution 3.1. *Proof.* Show that if $|x - x_0| < \delta$ then $|f(x) - L| < \varepsilon$.
If $|x - 2| < \delta$ then $|x^2 + x - 2 - 4| < \varepsilon$. Let $\delta = \frac{\varepsilon}{6}$ (Note 3.1). If $|x - 2| < \delta = \frac{\varepsilon}{6} \Rightarrow 6|x - 2| < \varepsilon \Rightarrow |x + 3||x - 2| < \varepsilon$

Note 3.1. $|x^2 + x - 2 - 4| < \varepsilon \Rightarrow |(x - 2)(x + 3| < \varepsilon \Rightarrow |x - 2| < \frac{\varepsilon}{x+3}$. Also, note that if $\delta = 1 \Rightarrow |x - 2| < \delta = 1 \Rightarrow -1 < x - 2 < 1 \Rightarrow 4 < x + 3 < 6$. Then $\delta = min\{\frac{\varepsilon}{6}, 1\}$.

Solution 3.2. *Proof.* Show that if $|x - x_0| < \delta$ then $|f(x) - L| < \varepsilon$.
If $|x - 2| < \delta$ then $|\frac{1}{x} - \frac{1}{2}| < \varepsilon$. Let $\delta = 2\varepsilon$ (Note 3.2). If $|x - 2| < \delta = 2\varepsilon \Rightarrow \frac{|x-2|}{2} < \varepsilon$, but $\frac{|x-2|}{2|x|} < \frac{|x-2|}{2}$ then $\frac{|x-2|}{2|x|} < \varepsilon$

Note 3.2. $|\frac{1}{x} - \frac{1}{2}| < \varepsilon \Rightarrow |\frac{x-2}{2x}| < \varepsilon \Rightarrow \frac{|x-2|}{2} < |x|\varepsilon$. Also, note that if $\delta = 1 \Rightarrow |x - 2| < \delta = 1 \Rightarrow -1 < x - 2 < 1 \Rightarrow 1 < x < 3$. Then $\delta = min\{2\varepsilon, 1\}$.

Solution 3.3. *Proof.* Show that if $|x - x_0| < \delta$ then $|f(x) - L| < \varepsilon$ (Suppose L exists).
If $|x - 0| < \delta$ then $|\frac{1}{x} - L| < \varepsilon$. Let $\delta = \frac{L-1}{\varepsilon}$ (Note 3.3). If $|x| < \delta = \frac{L-1}{\varepsilon} \Rightarrow \frac{|L-1|}{|x|} > \varepsilon$, but there is no element L that satisfies that inequality. So the limit does not exist.

Note 3.3. $|\frac{1}{x} - L| < \varepsilon \Rightarrow |\frac{1-xL}{|x|}| < \varepsilon \Rightarrow \frac{|1-Lx|}{\varepsilon} < |x|$. Also, note that if $\delta = 1 \Rightarrow |x| < 1 \Rightarrow L - 1 < Lx - 1 < L - 1$. Then $\delta = min\{\frac{L-1}{\varepsilon}, 1\}$.

Solution 3.4. (i) The graph of the function g (Fig. **3.9**) has the domain restricted to $\mathbb{R}^+$, so the graph of $f \circ g$ will only appear in $\mathbb{R}^+$ (Fig. **3.10**). (ii) The limit $\lim_{x \to 3} f \circ g = f[\lim_{x \to 3} g(x)] = f(\sqrt[4]{3}) = [\sqrt[4]{3}]^3 = 2.28$. Then the composite function $f \circ g: \mathbb{R}^+ \to \mathbb{R}, \ x^{\frac{3}{4}}$.

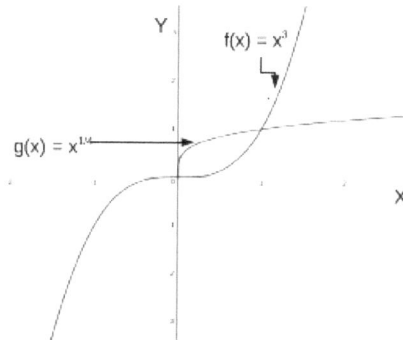

Fig. (3.9). Graphic description of the functions f and g.

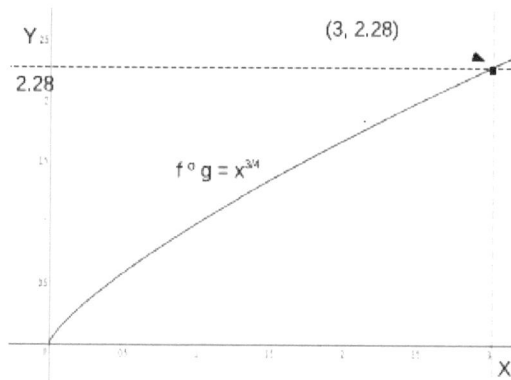

Fig. (3.10). Graphic description of the function $f \circ g$.

Note 3.4. Note that $\lim_{x \to \infty} f \circ g = \infty$.

Solution 3.5. Be $f(x) = [x]$, where $[x]$ means the greatest integer less than or equal to x. Be $x \to 3^+$

Table 3.1. Evaluation of $f(x) = [x]$.

x	3.1	3.01	3.001	3.0001	3.00001	3.000001
$f(x) = [x]$	3	3	3	3	3	3

Then $\lim_{x \to 3^+} [x] = 3$. $x \to 2$. The values of $f(x)$ at points less than 3 and close to 3 are noted (Table **3.5**).

Table 3.2. Evaluation of $f(x) = [x]$.

x	2.9	2.99	2.999	2.9999	2.99999	2.999999
$f(x) = [x]$	2	2	2	2	2	2

Therefore $\lim_{x \to 3^-}[x] = 2$. Since the **left hand limit** ($\lim_{x \to 3^+}[x] = 3$) and the **right hand limit** ($\lim_{x \to 3^-}[x] = 2$) are not equal, the $\lim_{x \to 3}[x]$ does **not** exist.

Solution 3.6. (Solution taken from [42]**).** (i) See (Fig. **3.11**)

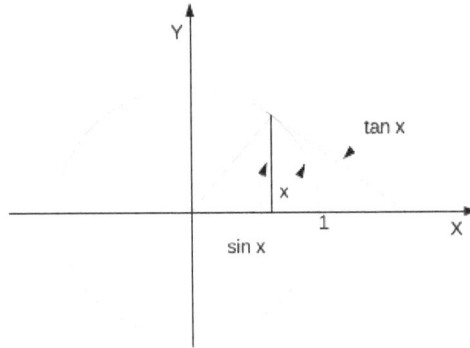

Fig. (3.11). Trigonometric relation of $\sin x$, x, and $\tan x$.

(ii)

$$\sin x < x < \tan x \tag{3.8}$$

Since $x \to 0 \Rightarrow \sin x \neq 0$

$$\frac{\sin x}{\sin x} < \frac{x}{\sin x} < \frac{\tan x}{\sin x} \tag{3.9}$$

then

$$1 < \frac{x}{\sin x} < \frac{\sin x}{\cos x \sin x} = \frac{1}{\cos x} \tag{3.10}$$

$$\lim_{x \to 0} \frac{1}{\cos x} = 1$$

$$\therefore \lim_{x \to 0} \frac{x}{\sin x} = 1 \tag{3.11}$$

$$\therefore \lim_{x\to 0}\frac{\sin x}{x}=1 \tag{3.12}$$

Solution 3.7.

$$\lim_{x\to\infty}\frac{\sqrt[3]{27x^3-2x+5}}{x+1}=\lim_{x\to\infty}\frac{\sqrt[3]{27x^3}}{x}$$

$$=\lim_{x\to\infty}\frac{3x}{x} \tag{3.13}$$

$$=3$$

See (Fig. **3.12**)

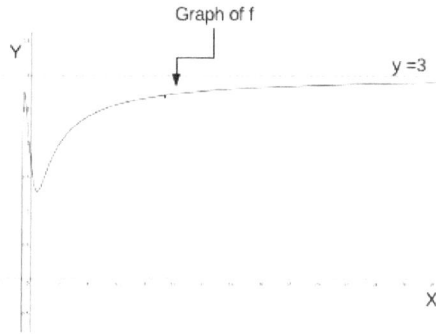

Graph of f

y =3

Fig. (3.12). Graph of the function $\frac{\sqrt[3]{27x^3-2x+5}}{x+1}$.

$$\lim_{x\to 1^-}\frac{\sqrt[3]{x^2+2x-3}}{|x-1|}=\lim_{x\to 1^-}\frac{(x-1)(x+3)}{-(x-1)}$$

$$=\lim_{x\to 1^-}-x-3 \tag{3.14}$$

$$=-4$$

and,

$$\lim_{x \to 1^+} \frac{\sqrt[3]{x^2+2x-3}}{|x-1|} = \lim_{x \to 1^+} \frac{(x-1)(x+3)}{(x-1)}$$

$$= \lim_{x \to 1^+} x + 3 \tag{3.15}$$

$$= 4$$

Both limits (Eqs. 3.14, 3.15) are not equal. The limit does not exist at "1".

See (Fig. **3.13**)

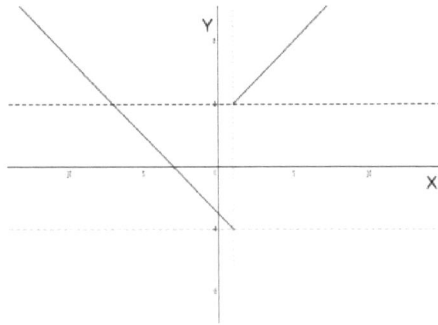

Fig. (3.13). Graph of the function $\frac{x^2+2x-3}{x-1}$.

Solution 3.9. Factor x^2 in the numerator and the denominator and simplify.

$$\lim_{x \to \infty} \frac{x-1}{2x^2+3} = \lim_{x \to \infty} \frac{\frac{1}{x}-\frac{1}{x^2}}{2+\frac{3}{x^2}}$$

$$= \lim_{x \to \infty} \frac{0}{2} \tag{3.16}$$

$$= 0$$

See (Fig. **3.14**)

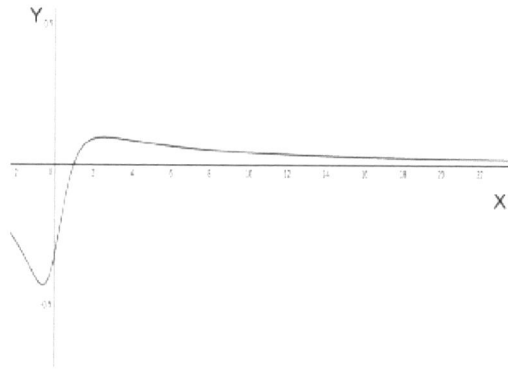

Fig. (3.14). Graph of the function $\frac{x-1}{2x^2+3}$.

Solution 3.10. Let $z = \frac{1}{x}$, as x gets large x approaches 0. Substitute and calculate the limit as follows.

$$\lim_{x \to \infty} x \sin \frac{1}{x} = \lim_{z \to 0} \frac{\sin z}{z}$$

$$= \quad 1 \qquad\qquad \textbf{(3.17)}$$

See (Fig. **3.15**)

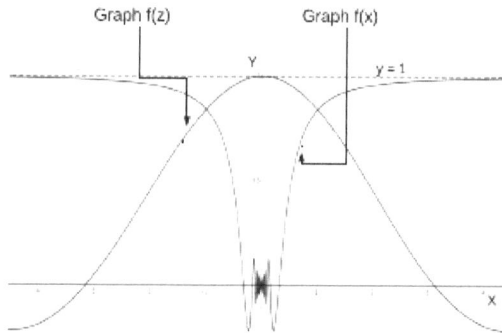

Fig. (3.15). Graph of the function $f(x) = x \sin \frac{1}{x}$, and its transformation $f(z) = \frac{\sin z}{z}$.

Chapter 4

Solution 4.1. When a continuous function is delimited in its domain and its graph cannot increase or decrease indefinitely, then the function f will acquire a maximum or minimum value at some point in its dimensioned domain. If the function represents a constant value, then any element of the domain will be maximum or minimum.

Solution 4.2. Maps (Def. 2.2) do not have the restriction that each element in their domain must correspond to each element of its **image**. Therefore, graphs like (Fig. 4.5) can be defined with a **map**. So it can be stated that any **function** f is contained in the set of **maps**.

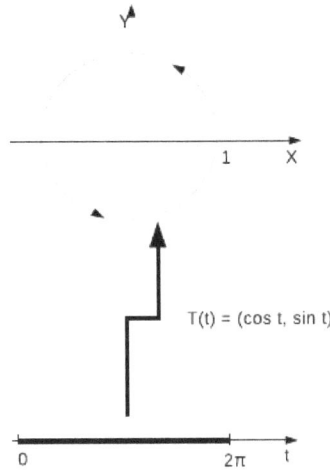

Fig. (4.5). Graphic description of the map T.

Solution 4.3. When a continuous function has at the limits of its domain a positive and a negative value (or inversely), then the graph necessarily has to cross the x-axis at point x_0, thus $f(x_0) = 0$. Although it does not imply a change in concavity.

Solution 4.4. $\lim_{x\to\infty} \frac{x+1}{x^4+2x+2} = \lim_{x\to\infty} \frac{\frac{x}{x^4}+\frac{1}{x^4}}{\frac{x^4}{x^4}+\frac{2x}{x^4}+\frac{2}{x^4}} = \frac{0}{1} = 0.$

Solution 4.5. Since function f is continuous, every element of its image D_f comes from an element of its domain D_f. Therefore, $\lim \lim_{x\to x_0 f(x)=L}$.

Solution 4.6. Since $x^3 - 4x^2 + 3x = x(x-1)(x-3) = 0$, then f has three indetermination points a–$\mathbb{R} = \{0,1,3\}$. Thus, we gather that g is continuous in A as f is, and $f(x) = g(x)$ si $x \in A$. If $x \notin A$ we have three alternatives.

Case $x = 0$:
$$\lim_{x \to 0^+} g(x) = \lim_{x \to 0^+} \frac{(x-1)|x-3|}{x^3 - 4x^2 + 3x} e^{1-x} = +\infty$$

$$\lim_{x \to 0^-} g(x) = \lim_{x \to 0^-} \frac{(x-1)|x-3|}{x^3 - 4x^2 + 3x} e^{1-x} = -\infty.$$

Thus, the limit of g does not exists, and g is not continuous in 0.

Case $x = 1$:
$$\lim_{x \to 1} g(x) = \lim_{x \to 1} \frac{(x-1)|x-3|}{x^3 - 4x^2 + 3x} e^{1-x} = -1.$$

Thus, the limit of g is different than $g(1) = 0$, therefore g is not continuous in 1.

Case $x = 3$:
$$\lim_{x \to 3^+} g(x) = \lim_{x \to 3^+} \frac{(x-1)|x-3|}{x^3 - 4x^2 + 3x} e^{1-x} = \frac{e^{-2}}{3}$$

$$\lim_{x \to 3^-} g(x) = \lim_{x \to 3^-} \frac{(x-1)|x-3|}{x^3 - 4x^2 + 3x} e^{1-x} = -\frac{e^{-2}}{3}.$$

Thus, both limits are different so the limit of g does not exist, therefore g is not continuous in 3.

Solution 4.7. A function f is continuous in x_0 when, for any boundary of $f(x_0)$ that we set, you can find a boundary of x_0 whose corresponding images I_f are contained in the boundary of $f(x_0)$. This implies: (i) $\exists f(x_0)$ and (ii) $\forall\, \varepsilon > 0\ \exists\, \delta > 0$ such that $|x -x_0| < \delta$ implies $|f(x) - f(x_0)| < \varepsilon$.

Solution 4.8. f is continuous when $x \neq 0$; at $x = 0$ we have $\lim_{x\to 0} f(x) = 0$, $f(0) = 1$. Then f is continuous in $\mathbb{R}$.

Solution 4.9. (i) $f \circ g = f(g) = \cos x^2 + 1$ is continuous in $\mathbb{R}$. (ii) $g \circ f = g(f) = \cos x^2 + 1$ is continuous in $\mathbb{R}$.

Solution 4.10. The function $\frac{\log 1+x^2}{x^4-26x^2+25}$ has discontinuities in $-5, -5, -1, 1$, then a valid domain to be continuous is $[6,10]$.

Chapter 5

Solution 5.1. (i) $f'(x) = \lim_{h\to 0} \frac{f(x+h)-f(x)}{h} = \lim_{h\to 0} \frac{(x+h)^2-x^2}{h} = \lim_{h\to 0} \frac{h(2x+h)}{h}$
$= \lim_{h\to 0} 2x = 2x$. (ii) $y - y_0 = f'(3)(x - x_0) \Leftarrow y - 9 = 6(x - 3) \Leftarrow y - 9 =$
$6x - 18 \Leftarrow y = 6x - 9$ substituting point $x = 3$ we have $y = 9$, therefore, $(3,9)$
is a common point in the graph of $f(x) = x^2$ and the line $y = 6x - 9$. (iii) See
(Fig. **5.7**).

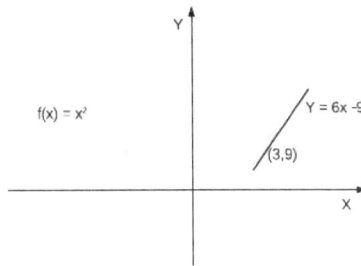

Fig. (5.7). Graph of the tangent line at point 3.

(iv) The function f is polynomial then $f'(x) = 2x$.

Solution 5.2. (i) If $f(x) = x^2 - 2x + 3$ then the critical points will be located
where $f'(x) = 0 \Leftarrow 2x - 2 = 0 \Leftarrow x = 1$. There is only one critical point $x = 1$.
(ii) $f''(x) = 2 > 0$ then, the function has a minimum at point $x = 1$.

Solution 5.3. We have to minimize the function $f(x,y) = xy$, with the restriction
$x - y = 10$. So $y = x - 10$, substituting in f, we have $f(x) = x(x - 10) = x^2 -$
$10x$. The critical points are $f'(x) = 2x - 10 = 0 \Leftarrow x = 5$. Since $f''(x) = 2 > 0$,
then the function in $x = 5$ has the minimum values $x = 5$ and $y = -5$.

Solution 5.4. If $f(x) = x^3 - 3x + 1$, then $f'(c) = 3c^2 - 3$, so $3c^2 - 3 =$
$\frac{(27-9+1)-(1-3+1)}{2} = \frac{20}{2} = 10$. Thus, $3c^2 - 3 = 10 \Leftarrow c = \pm\sqrt{\frac{7}{3}}$. but $c \in (1,3)$, so
$c = \sqrt{\frac{7}{3}}$

Solution 5.5. We have to minimize the closed surface of the cylinder, that is minimizing the function $S(r,h) = 2\pi rh + 2\pi r^2$ with the restriction $V(r,h) = \pi r^2 h = 330ml \Leftarrow h = \frac{330}{2\pi r^2}$. Since r and h are defined in dm (decimeters) and $1l = 1dm^3$, then substituting h in the function S we have $S(r) = \frac{330}{r} + 2\pi r^2$. Now, to get the critical points of S we have $S'(r) = 0 \Leftarrow S'(r) = -\frac{330}{r^2} + 4\pi r = 0 \Leftarrow r = \sqrt[3]{\frac{165}{2\pi}}$. We compute $S''(r) = \frac{600 + 4\pi r^3}{r^3}$. Since the value of r is positive, then $S''(r) > 0$; so the critical point $r = \sqrt[3]{\frac{165}{2\pi}}$ is where S has a maximum value.

The dimensions of the minimum surface of the cylinder are $r = \sqrt[3]{\frac{165}{2\pi}} dm^3$, and $h = \frac{330}{2\pi(\frac{165}{2\pi})^{\frac{2}{3}}} = \frac{330}{\pi\sqrt[3]{\frac{165^2}{4\pi^2}}} dm^3$.

Solution 5.6. (i) $T'(t) = (-\sin t, \cos t)$ then $T'(\pi) = (0, -1)$. (ii) It is a line on point $T(\pi) = (\cos \pi, \sin \pi) = (-1, 0)$ directed to vector $T'(\pi) = (0, -1)$. The Cartesian expression for the tangent line is $x = -1$. (iii) See (Fig. **5.8**).

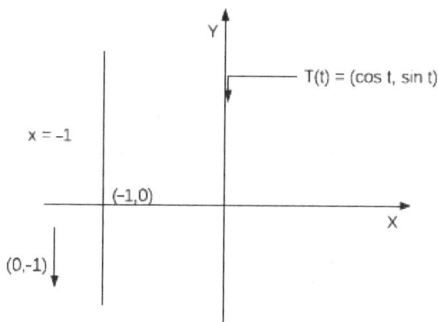

Fig. (5.8). Graph of the tangent line at point π.

Solution 5.7. $\lim_{x \to 3} \frac{1}{x-3} - \frac{5}{x^2-x-6} = \lim_{x \to 3} \frac{x-3}{x^2-x-6} = \lim_{x \to 3} \frac{1}{2x-1} = \frac{1}{5}$.

Solution 5.8. Using implicit differentiation (i) $6x - 2y = 0 \Leftarrow 6 - 2\frac{dy}{dx} = 0 \Leftarrow \frac{dy}{dx} = 3$. (ii) $\sec^2 x + \csc^2 y = 0$, then $\frac{dy}{dx} = -\frac{\frac{\partial G}{\partial x}}{\frac{\partial G}{\partial y}}$, where $G(x,y) = \sec^2 x + \csc^2 y$, so $\frac{dy}{dx} = -\frac{\sec^2 x \tan x}{\csc^2 y \cot y}$.

Solution 5.9. (i) $G(x, y) = 6x - 2y$ implies $\frac{\partial G}{\partial y} = 2$. Since $2 \neq 0$ then $\forall y \in \mathbb{R}$ is an invertible function. (ii) $G(x, y) = \sec^2 x + \csc^2 y$ implies $\frac{\partial G}{\partial y} = \csc^2 y \cot y = \frac{\cos y}{\sin^3 y}$, so the function is invertible when $\sin y \neq 0$ *i.e.* $y \neq n\pi$, $n \in \mathbb{Z}$.

Solution 5.10. *Proof.* (Proof taken and adapted from [71]) If $x^n = e^{n \ln x}$ is applied ln in both sides, then $(x^n)' = (e^{n \ln x})' = \frac{n}{x} e^{n \ln x} = \frac{n}{x} x^n = n x^{n-1}$.

Chapter 6

Solution 6.1. (i) $1,3,5,7,9,\cdots$. (ii) $3,-3,3,-3,\cdots$. (iii) $2,-3,2,-4,2,\ldots$. (iv) It is not possible. (v) It is not possible. (vi) $-6,6,6,6,\cdots$. (vii) $0.2,-0.2,0.002,-0.002$,$0.0003,-0.0003,\cdots$. (viii) $6,3,7,4,8,5,\cdots$.

Solution 6.2. Taking the difference $a_n - a_{n-1}$ of the sequence $8,3,-2,-7,-12$, $3-8=-5$, $7-2-3=-5$, $-7-(-2)=-5$ and $-12+7=-5$, then the difference $d=5$. So the **general term** is $a_n = 8 + (n-1)d = 8 - 5(n-1) = 8 - 5n + 5 = 13 - 5n$, i.e. $a_n = 13 - 5n$ (Note 6.5).

Note 6.5. The general term of an arithmetic sequence is $a_n = a_k + (n-k)d$.

Solution 6.3. The sequence (a) is strictly monotonically increasing and it is bounded in the interval $[2,3]$, therefore, it is convergent to the irrational number e which is its limit (Table **6.3**).

$$\lim_{n\to\infty}\left(1+\frac{x}{n}\right)^n \text{ (Note 6.6)} = \lim_{n\to\infty}e^{n\ln(1+\frac{x}{n})} = e^{\lim_{n\to\infty}n\ln(1+\frac{x}{n})} =$$

$$e^{\lim_{n\to\infty}\frac{\ln(1+\frac{x}{n})}{\frac{1}{n}}}.$$

Note 6.6. $e^{\ln(1+\frac{x}{n})^n} = e^{n\ln(1+\frac{x}{n})}$

Applying (Sect. 5.2): $e^{\lim_{n\to\infty}\frac{-(\frac{x}{n^2})(1+\frac{x}{n})}{-\frac{1}{n^2}}} = e^{\lim_{n\to\infty}\frac{x}{1+\frac{x}{n}}=e^x}$, therefore, $(1+\frac{x}{n})^n \to e^x$ [100].

Table 6.3. Sequence (a_n).

n	1	2	3	4	5	6	7
$\left(1+\frac{x}{n}\right)^n$	2	2.25	370371	2.441406	2.488320	2.521626	2.546501

Solution 6.4. The difference of two successive terms of the sequence $\{0,3,8,15,$ $24,35\}$ is $d_1 = 3-0 = 3$, $d_2 = 5$, $d_3 = 7$, $d_4 = 9$ and $d_4 = 11$, then $a_1 = 0$, $a_2 = 3$ and $a_n = (2n-1) + n$ (Note 76).

Solution 6.5. No, it is not because it is not bounded and is strictly monotonically increasing. This is not a Cauchy sequence (Def. 6.4), because for a value $\varepsilon > 0$ there exists only a finite number of terms that meet $m, n > N \Rightarrow |a_n - a_m| < \varepsilon$ and an infinite number of terms that do not meet that.

Solution 6.6. A recursive sequence is one whose terms are determined from the value of the previous terms i.e. $(a_n) = a_{n-1}$, where $\alpha \in \mathbb{R}$ *e.g.* $a_1 = 2$, its general term is $a_n = 3a_{n-1}$, so $(a_n) = \{2, 6, 18, \cdots, 3a_{n-1}\}$.

Solution 6.7. The representation in $\mathbb{R}^2$ shows the conversion of the sequence in the Y −axis that represents the image of the sequence, whilst the representation in $\mathbb{R}$, is the representation itself. Note that the procedure of representing a sequence in a plane or in the real line is different.

Solution 6.8. No, it is not because there is not any number $n \in \mathbb{N}$, such that $3n + 12 = 25$.

Solution 6.9. $(a_n) = \{2, -4, 8, -16, \cdots\}$. The sequence is not monotonically, it is not convergent, it is not divergent, and it is not bounded.

Solution 6.10. Since the sequence (a_n) is convergent, then the subsequence is convergent.

Chapter 7

Solution 7.1. $\lim_{n \to \infty} |\frac{a_{n+1}}{a_n}| = \lim_{n \to \infty} |\frac{\frac{1}{2^{n+1}}}{\frac{1}{2^n}}| = \frac{\frac{1}{2^n 2}}{\frac{1}{2^n}} = \frac{1}{2} < 1.$ The series is **convergent**.

Note that the value $\frac{1}{2}$ **is not** the value the series converges to, but the limit the method uses to determine whether the series converges or not.

In order to know the value the series converges to, it is necessary to determine the **sequence of partial sums** a_S, assuming that the limit of a_S is the value the series converges to.

The first terms of the series are: $\{\frac{1}{2^0}, \frac{1}{2^1}, \frac{1}{2^2}, \cdots\}$ that have the form ?, $\sum_{k=1}^{n-1} ar^k$, where $a = 1$ and $r = \frac{1}{2}$. This sum can be expressed $s = ar + ar^2 + ar^3 + \cdots + ar^{n-1}$ that implies $rs = ar + ar^2 + ar^3 + \cdots + ar^n$ and $s - rs = a - ar^n$, where $s = a(\frac{1-r^n}{1-r})$, $r \neq 1$. This term is the **sequence of partial sums** of the series, so substituting $a = 1$ and $r = \frac{1}{2}$ we have $\lim_{n \to \infty} 2 - \frac{1}{2^{n-1}} = 2$, thus, the value of the series is 2.

Solution 7.2. $\lim_{n \to \infty} |\frac{a_{n+1}}{a_n}| = \lim_{n \to \infty} \frac{n^n}{(n+1)^n} = \lim_{n \to \infty} \frac{1}{(1+\frac{1}{n})^n} = \frac{1}{e} > 1.$ The series is convergent.

Solution 7.3. Let $\sum_{n=1}^{\infty} a_n$ and $\sum_{n=1}^{\infty} b_n$, two series of positive terms, such that $a_n \leq b_n, \forall n \in \mathbb{N}$ [112], so

1. If the series $\sum_{n=1}^{\infty} b_n$ is **convergent**, then the series $\sum_{n=1}^{\infty} a_n$ is **convergent** too.

2. If the series $\sum_{n=1}^{\infty} a_n$ is **divergent**, then the series $\sum_{n=1}^{\infty} b_n$ is **divergent** too.

Note 7.7. Is the series $\sum_{n=1}^{\infty} \frac{\sin^2 2n}{3^n}$ convergent? Since $\frac{\sin^2 2n}{3^n} < \frac{1}{3^n}, \forall n \in \mathbb{N}$ and the series $\sum_{n=1}^{\infty} \frac{1}{3^n}$, then the series $\sum_{n=1}^{\infty} \frac{\sin^2 2n}{3^n}$ is **convergent**.

Solution 7.4. Let $\frac{1}{3} = 0.3333 \cdots = \frac{3}{10} + \frac{3}{100} + \frac{3}{1000} + \cdots + \frac{3}{10^n}$, the sequence of partial sums a_S is determined by $a_{S=1} = \frac{3}{10} = \frac{1}{3}\left(1 - \frac{1}{10}\right)$, $a_{S=2} = \frac{3}{10} + \frac{3}{100} = \frac{33}{10^2} = \frac{1}{3}\left(1 - \frac{1}{10^2}\right)$ and so on $a_{S=n} = \frac{3}{10} + \frac{3}{100} + \cdots + \frac{3}{10^n} = \frac{33\overline{33}}{10^n} = \frac{1}{3}\left(1 - \frac{1}{10^n}\right)$. $\lim_{n\to\infty} \frac{1}{3}\left(1 - \frac{1}{10^n}\right) = \frac{1}{3}$, so $\sum_{n=1}^{n} \frac{3}{10^n} = \frac{1}{3}$.

Remark 7.1. Note that the convergence of the series was not verified, as what was requested was the value of the series. Usually it is necessary to verify whether the series is convergent or not.

Solution 7.5. If $\frac{2n+1}{n^2(n+1)^2} = \frac{1}{n^2} - \frac{1}{(n+1)^2}$, then $\sum_{n=0}^{\infty} \frac{2n+1}{n^2(n+1)^2} = \left(1 - \frac{1}{4}\right) + \left(\frac{1}{4} - \frac{1}{9}\right) + \left(\frac{1}{9} - \frac{1}{16}\right) + \cdots + \left(\frac{1}{n^2} - \frac{1}{(n+1)^2}\right)$. Note that the middle cancell each other, therefore, the **sequence of partial sums** is $a_S = 1 - \frac{1}{(n+1)^2}$. $\lim_{n\to\infty} 1 - \frac{1}{(n+1)^2} = 1$. The series is **convergent** and its value is 1.

Solution 7.5. The Taylor polynomial becomes the Maclaurin polynomial when the value of x tends to 0.

Solution 7.7. Given the importance of both theorems and their relation with the Taylor series, here we reproduce the comment (Nt. 5.9) mentioned in (Chap. 5). "When these theorems are reviewed for the first time, we get the impression that something is missing. The Implicit Function Theorem gives the conditions to explicitly know the derivative f' of a function f, it can even give us the subset of the domain of function f where its inverse f^{-1} exists, however, we don't know function f. On the other hand, the Inverse Function Theorem is a tool to know the derivative of the inverse function $(f^{-1})'$ from the derivative f' of function f, but it doesn't let us know function f, or its inverse function f^{-1}.

In summary, both theorems let us know explicitly the derivatives f' of function f and of its inverse $(f^{-1})'$, but they don't let us know the functions f or f^{-1}. When we review Taylor's series (Ch. 7), we will find that from the derivative f' of a function f, we can determine function f, the same when using the derivative of the inverse function $(f^{-1})'$ we can determine the function f^{-1}.

These theorems and Taylor's series make possible to connect two spaces from a subset that is a **bijection**. First knowing the derivatives and then, from Taylor's series, knowing the functions. The Implicit and Inverse Function Theorems are fundamental in Differential Calculus".

Solution 7.8. From Euler's formula $e^{ix} = \cos x + i\sin x = -1 \Rightarrow e^{i\pi} + 1 = 0 \Rightarrow$ $(e^{i\pi})^n = (-1)^n$. Thus $e^{ix} = \cos x + i\sin x = -1$ when $x = n\pi$ $e^{in\pi} = \cos n\pi +$ $i\sin n\pi = (-1)^n$. This implies that $\cos n\pi = (-1)^n$ and $\sin n\pi = 0$ for all $n \in \mathbb{Z}$ (**Proof taken from** [113]).

$$\sum_{n=2}^{\infty} \frac{\cos n\pi}{\sqrt{n}} = \sum_{n=2}^{\infty} \frac{(-1)^n}{\sqrt{n}} \Rightarrow a_n = \frac{1}{\sqrt{n}}$$

Since (Def. 7.5) (a_n) is decreasing and $\lim_{n\to\infty} \frac{1}{\sqrt{n}=0}$, the series is **convergent**.

Solution 7.9. By (Eq. 7.3) $\lim_{n\to\infty}|\sqrt[n]{|(\frac{n^2+1}{2n^2+1})^n|}| = \lim_{n\to\infty} \frac{n^2+1}{2n^2+1} = \frac{1}{2} < 1$. If $|x| < \frac{1}{2}$, then the series converges.

Solution 7.10. The **sequence of partial sums** starting $n = 0$? is $a_S = \{1, 1 + \frac{1}{2}, 1 + \frac{1}{2} + \frac{1}{4}, 1 + \frac{1}{2} + \frac{1}{4} + \frac{1}{8}\}, \cdots$, from that representation $a_{S=0} = 2 - 1$, $a_{S=1} = 2 - \frac{1}{2}$, $a_{S=2} = 2 - \frac{1}{4}$, $a_{S=3} = 2 - \frac{1}{8}$, which means that the general term is $a_S = 2 - \frac{1}{2^{n-1}}$. Since $\lim_{n\to\infty} a_S = \lim_{n\to\infty} 2 - \frac{1}{2^{n-1}} = 2$, the series is **convergent**.

Chapter 8

Solution 8.1. $(f_k) = \{1,2,3,4,\cdots,k\}$.

Solution 8.2. $(f_k)_{[0,1]} = x^k = \{1, x, x^2, x^3, \cdots, x^n\}$, $(f_k)_{(1,\infty)} = 1 = \{1,1,1,1,\cdots,1\}$, and
$\lim_{k\to\infty} f_k(x) = \begin{cases} 0 & for & x \in [0,1] \\ 1 & for & x \in (1,\infty) \end{cases}$

Solution 8.3. $f_k(0) = f(0) = 1$ and $x \in (0,1]$, $|f_k(x) - f(x)| = |f_k(x)| = 0 < \varepsilon$. If $k > \frac{1}{x}$, then the sequence of functions $f_k(x)$ converges pointwise to $f(x)$.

Solution 8.4. See (Fig. **8.7**).

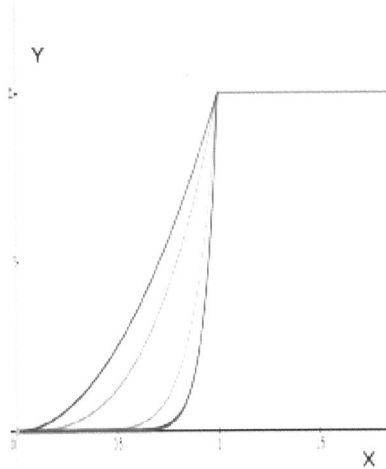

Fig. (8.7). Graph of (f_k).

Solution 8.5. No, the convergence of a sequence of functions is not related to the derivative of the sequence. Counterexample, let the sequence of functions $(f_k) = \frac{\sin kx}{k}$, its $(f'_k(x)) = \cos kx$ does not converge pointwise.

Solution 8.6. $(x_j e^{-x_j})_{j=1,3,5}^{n=1} = \{(1e^{-1})^1, (3e^{-3})^1, (5e^{-5})^1\}$.

Solution 8.7. See (Fig. **8.8**).

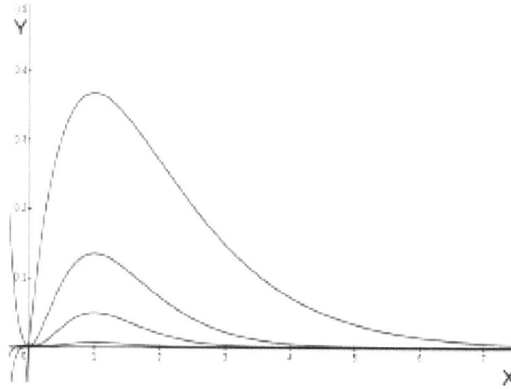

Fig. (8.8). Graph of (f_k).

Solution 8.8. If the sequence of functions is uniform convergent, then it is pointwise convergent and not the other way round.

Solution 8.9. Yes, Weierstrass (1872) proposes the series of functions $\sum_{i=1}^{n} b^n \cos(a^n x)$, which is uniform convergent for $b < 1$ and does not have derivative at any point in its domain for $ab > 1 + \frac{3\pi}{2}$ [92].

Solution 8.10. $\sum_{n=1}^{n} (xe^{-x})^n = (x_0 e^{-x_0})^1, (x_0 e^{-x_0})^1 + (x_0 e^{-x_0})^2, (x_0 e^{-x_0})^1 + (x_0 e^{-x_0})^2 + (x_0 e^{-x_0})^3, \cdots.$

Chapter 9

Solution 9.1. Since $F'(x) = \sin 2x$, then $F(x) = \int f(x)\ dx = \int \sin 2x\ dx = \frac{1}{2}\int \sin u\ du = -\cos u + c$.

Solution 9.2. Note that the function is negative in the interval $[-1,0]$, then $area = |\int_{-1}^0 x\ dx| + \int_0^1 x\ dx = \frac{1}{2} + \frac{1}{2} = 1$.

Solution 9.3. Yes, it is. From (Thm. 3) $\int_0^\pi f(x)\ dx = F(\pi) - F(0) = b - a$.

Solution 9.4. $\int_1^2 xe^{x^2}\ dx$ ($u = x^2$, $du = 2xdx \Rightarrow dx = \frac{du}{2x}$, $u_1 = 1$ and $u_2 = 4$), then $\int_1^2 xe^{x^2}\ dx = \frac{1}{2}\int_1^4 e^u\ du = \frac{1}{2}(e^4 - e)$.

Solution 9.5. $\int x^2\cos x\ dx = x^2\sin x - [\int 2x\sin x\ dx] = x^2\sin x - [-x\sin x + \cos x] = (x^2 - 2)\sin x + 2x\cos x + c$.

Solution 9.6. $\int_0^1 \frac{1}{2+x}\ dx$ If $u = 2 + x$, $du = dx$, $u_1 = 2$, and $u_2 = 3$, then $\int_0^1 \frac{1}{2+x}\ dx = \ln 3 - \ln 2$.

Solution 9.7. From the geometrical point of view, both definite integrals represent the same area in different intervals and different graphs.

Solution 9.8. $\int_0^1 \frac{1}{\sqrt{x}}\ dx = \lim_{a\to 0}\int_a^1 \frac{1}{\sqrt{x}} = \lim_{a\to 0}[2\sqrt{x}]_a^1 = \lim_{a\to 0}2 - 2\sqrt{a} = 2$.

Solution 9.9. $J(T(u)) = \left|\frac{dT}{du}\right| = e^u(u^3 + 3u^2)$.

Solution 9.10. (i) $\int_0^2 x^3\ dx = \left[\frac{1}{4}x^4\right]_0^2 = 4$. (ii) Let $T(u) = u^{\frac{1}{3}}, u \in [0,8]$. (iii) If $T(u) = u^{\frac{1}{3}}, J(T(u) = \frac{1}{3}u^{-\frac{2}{3}}, \int_0^2 x^3\ dx = \int_0^8 f\circ T(u)|J(T(u)|du = \frac{1}{3}\int_0^8 u^{\frac{1}{3}}\ du = 4$.

Further Reading

The following articles can serve the reader to understand the application of mathematics, in the same discipline, although their level is beyond a non-graduate student, A. Akgul [97–108].

The Paul's online notes describes the main operators, under simple language, and helps with graphs [64], A. Wade [18, 88], and P. Shunmugaraj [78].

The following textbooks were written for the level of non-graduates, and show a language accessible to the reader, H. Arizmendi-Peimbert, A.M. Carrillo-Hoyo and M. Lara-Aparicio [25], J.E. Marsden and A. Tromba [62], D. Varberg and EJ. Purcell and SE. Rigdon [109], and S. K. Chung [34].

This work contains a large number of useful exercises, the solutions of which have been reviewed, F. Ayres Jr. and E. Mendelson [81].

The following textbooks have greater language and rigor than desirable, however, they are still recommended for consultation, C. Polanco [96], M. Spivak [33], and R.G. Bartle and D.R. Sherbert [20].

Carlos Polanco
All rights reserved-© 2020 Bentham Science Publishers

References

[1]   Matesfacil, "Valor absoluto e inecuaciones," 2019. [Online]. Available:
       https://www.matesfacil.com/BAC/absoluto/valor-absolutoinecuacionesejercicios-resueltos.html

[2]   "N´umeros irracionales," 2019. [Online]. Available:
       https://numerosirracionales.com/operaciones-con-numeros-irracionales

[3]   "Descartes 2d, representaci´on gr´afica de los n´umeros: N´umeros irracionales, bajo licencia creative commons
       (cc) reconocimiento-compartirigual 3.0 espa˜na (cc by-sa 3.0 es)," 2019. [Online]. Available:
       http://recursostic.educacion.es/descartes/web/materiales didacticos/Representation en la recta/Numeros3.htm

[4]   K. E. Stange, "A function is bijective if and only if has an inverse," 2015. [Online]. Available:
       http://math.colorado.edu/ kstange/has-inverse-isbijective. pdf

[5]   "De entre todos lo tri´angulos rect´angulos de hipotenusa 4, determinarlas dimensiones del de ´area m´axima,"
       2012. [Online]. Available: https://www.profesor10demates.com/2012/11/ejercicios-yproblemas-resuelt os-de
       28.html

[6]   Matesfacil, "Problemas sobre funciones," 2019. [Online]. Available:
       https://www.matesfacil.com/ESO/funciones/problemas-resueltosfunciones-concepto-dominio-codominio-imagen-
       grafica.html

[7]   CENGAGE, "Limits and introduction to calculus," 2019. [Online]. Available:
       https://www.cengage.com/resource uploads/downloads/1111427631 267947.pdf

[8]   "Universidad nacional del litoral," 2019. [Online]. Available:
       https://www.fca.unl.edu.ar/Limite/Problemas%20Aplicaci%F3n%20cap2%20L%EDmite.htm

[9]   B. Cattaneo, "Matem´atica," 2019. [Online]. Available:
       https://rephip.unr.edu.ar/bitstream/handle/2133/5085/1407˙15%20MATEMATICA%20Continuidad.pdf?sequence
       =2&isAllowed=y

[10]   "Calcular las as´ıntotas de una funci´on," 2019. [Online]. Available:
       https://www.gaussianos.com/calcular-las-asintotas-de-una-funcion/

[11]   C.-C. Wang, "Continuity," 2018. [Online]. Available:
       http://web.ntpu.edu.tw/ ccw/calculus/Chapter 01/Page61-71.pdf

[12]   "Math open reference," 2011. [Online]. Available:
       https://www.mathopenref.com/calcboxproblem.html

[13]   "Siyavula," 2011. [Online]. Available:
       https://www.siyavula.com/read/maths/grade-12/differential-calculus/06-differential-calculus-07

[14]   D. Guichard, "Calculus: early transcendentals," 2019. [Online]. Available:
       https://www.whitman.edu/mathematics/calculus online/section06.02.html

[15]   Superprof, "L´ımites de sucesiones," 2019. [Online]. Available:
       https://www.superprof.es/apuntes/escolar/matematicas/aritmetica/sucesiones/limites-de-sucesiones-3.html

[16]   B. Saunders, "Chapter 2, limits of sequences," 2019. [Online]. Available: http://homepages.math.uic.edu/
       saunders/-MATH313/INRA/INRA Chapter2.pdf

[17]   C. R. Bermejo, "An´alisis matem´atico," 2019. [Online]. Available: http://www.mat.ucm.es/
       cruizb/MMI/Apuntes-i/Apuntes-16/Sucesiones-4.pdf

[18]   A. Wade, "Math 401 - notes. sequences of functions. Pointwise and uniform convergence," 2005. [Online].
       Available: http://www.personal.psu.edu/auw4/M401-lecture-notes.pdf

[19]   T. Vogel, "Pointwise and uniform convergence of sequences of functions (7.1)," 2019. [Online]. Available:
       https://www.math.tamu.edu/ tvogel/410/sect71a.pdf

[20]   R. Bartle and D. Sherbert, *Introduction to Real Analysis*. John & Wiley Sons, Inc., USA.

[21]   "The density of the rational numbers," 2001. [Online]. Available:
       http://mathforum.org/library/drmath/view/56025.html

[22]   Wikipedia, "Rational numbers." 2019. [Online]. Available:
       https://en.wikipedia.org/wiki/Rational number#Irreducible fraction

[23]   "Prove or disprove that the sum of two irrational numbers is irrational," 2019. [Online]. Available:
       https://math.stackexchange.com/questions/1415735/prove-or-disprovethat-the-sum-of-two-irrational-numbers-is-
       irrational

[24]   MathOnline, "The density of the rational/irrational numbers," 2015. [Online]. Available:
       http://mathonline.wikidot.com/the-density-of-the-rationalirrational-numbers

Carlos Polanco
All rights reserved-© 2020 Bentham Science Publishers

[25]   H. Arizmendi-Peimbert, A. Carrillo-Hoyo, and M. Lara-Aparicio, *C'alculo Primer Curso, Nivel Superior*. Instituto de Matem´aticas, Facultad de Ciencias, Universidad Nacional Aut´onoma de M´exico.

[26]   Wikipedia, "Triangle inequality," 2019. [Online]. Available: https://en.wikipedia.org/wiki/Triangle inequality

[27]   "Yumpu: Los n´umeros reales," 2019. [Online]. Available: https://www.yumpu.com/es/document/read/15126325/capitulo-1-losnumeros-reales-cimat/3

[28]   "Maths mutt mathematical resources, differentiating explicit and implicit functions," 2019. [Online]. Available: http://www.mathsmutt.co.uk/files/impex.htm

[29]   C. Polanco, "Advanced calculus– fundamentals of mathematics," 2019. [Online]. Available: Bentham Science Publishers (In Press)

[30]   Wikipedia, "Inverse function," 2019. [Online]. Available: https://en.wikipedia.org/wiki/Inverse function

[31]   G. Osborne, "Differential and integral calculus with examples and applications," 1906. [Online]. Available: http://library.umac.mo/ebooks/b31290735.pdf

[32]   T. W. Ng., "Seminar on advanced topics in mathematics, solving polynomial equations," 2008. [Online]. Available: https://hkumath.hku.hk/ ntw/EMB(poly2008).pdf

[33]   M. Spivak, *C'alculo Infinitesimal*. Editorial Revert´e, Barcelona, Espa˜na.

[34]   S. K. Chung, *Understanding Basic Calculus*, 2007. [Online]. Available: http://www.math.nagoya-u.ac.jp/ richard/teaching/f2016/BasicCalculus.pdf

[35]   P. Bhawalkar and K. Johnston, "Lumen boundless algebra, rational functions," 2019. [Online]. Available: https://courses.lumenlearning.com/boundlessalgebra/chapter/rational-functions/

[36]   "Desmos, inc.graphing calculator," 2019. [Online]. Available: https://www.desmos.com/calculator/v0rxp6hncb

[37]   "Zona land education, rational functions," 2019. [Online]. Available: http://zonalandeducation.com/mmts/functionInstitute/rationalFunctions/one OverX/oneOverX.html

[38]   "Sangaku maths, irrational functions," 2019. [Online]. Available: https://www.sangakoo.com/en/unit/irrational-functions

[39]   "L'ımite de una funci'on," 2008. [Online]. Available: https://definicion.de/limitede-una-funcion/

[40]   G. Vasco, "L'ımite de una funci'on," 2008. [Online]. Available: https://www.hiru.eus/es/matematicas/limite-de-una-funcion

[41]   Matesfacil, "Formal definition of epsilon-delta limits," 2019. [Online]. Available: https://brilliant.org/wiki/epsilon-delta-definition-of-a-limit/

[42]   "Math education and technology," 2003. [Online]. Available: http://www.iesmath.com/math/java/calc/LimSinX/LimSinX.html

[43]   Matesfacil, "C'alculo de l'ımites," 2019. [Online]. Available: https://www.matesfacil.com/BAC/limites/ejercicios-resueltos-limites-1.html

[44]   "Find limits of functions in calculus," 2019. [Online]. Available: https://www.analyzemath.com/calculus/limits/find limits functions.html

[45]   "Questions and answers on continuity of functions," 2019. [Online]. Available: https://www.analyzemath.com/calculus questions/continuity.html

[46]   "Paul's online urls," 2013. [Online]. Available: http://tutorial.math.lamar.edu/Problems/CalcI/Continuity.aspx

[47]   "Problems on continuity of functions," 2019. [Online]. Available: https://www.vitutor.com/calculus/limits/continuity problems.html

[48]   "Khan academy," 2019. [Online]. Available: https://www.khanacademy.org/math/ap-calculus-ab/ab-limits-new/ab-1-13/e/continuity

[49]   "Khan academy," 2019. [Online]. Available: https://www.khanacademy.org/math/ap-calculus-ab/ab-limits-new/ab-1-10/e/analyzing-discontinuities-graphical

[50]   "Wikipedia," 2019. [Online]. Available: https://en.wikipedia.org/wiki/Extreme value theorem

[51]   A. Bogomolny, "Cut the knot," 2018. [Online]. Available: https://www.cutthe-knot.org/fta/brodie.shtml

[52]   "Vitutor," 2019. [Online]. Available: https://www.vitutor.com/calculus/limits/Bolzano%27s%20Theorem.html

[53]   "Khan academy," 2018. [Online]. Available: https://es.khanacademy.org/math/differential-calculus/continuitydc/intermediate-value-theorem-dc/a/intermediate-value-theorem-review

[54]   "El blog del profe nelson," 2018. [Online]. Available: https://profbaptista.wordpress.com/2010/01/11/ejercicios-resueltoslimites-y-continuidad/

[55]   E. Cordero, "Continuity," 2018. [Online]. Available: http://webmath2.unito.it/paginepersonali/cordero/English/continuity.pdf

[56]   "Mathematics," 2018. [Online]. Available: https://www.embibe.com/engineering/practice/solve/jee/mathematics/differential-calculus/continuity-and-differentiability/session?start-sessionquestion=EM0076450

[57]   "Wikipedia," 2019. [Online]. Available: https://en.wikipedia.org/wiki/Derivative

[58]   BY'JUS, "Continuity and differentiability," 2019. [Online]. Available:
       https://byjus.com/maths/continuity-and-differentiability/
[59]   Wikipedia, "Leibniz's notation," 2019. [Online]. Available: https://en.wikipedia.org/wiki/Leibniz%27s notation
[60]   ——, "Lagrange's notation," 2019. [Online]. Available: https://en.wikipedia.org/wiki/Notation for
       differentiation#Lagrange's notation
[61]   J. M. Aguirregabiria, "Ecuaciones diferenciales ordinarias para estudiantes de f´ısica," 2000. [Online]. Available:
       https://webargitalpena.adm.ehu.es/pdf/UCWEB004524.pdf
[62]   J. Marsden and A. Tromba, *Vector Calculus*. New York, NY 10004, USA: WH Freeman And Company, 2011.
[63]   AWS, "Desarrollos en serie. series funcionales," 2019. [Online]. Available: https://s3.eu-west-
       1.amazonaws.com/eu.storage.safecreative.org/1/2011/07/20/00000131/4808/7f23/3870/5d4f173fa82d/Desarrollos
       enSerie.pdf?responsecontenttype=application%2Fpdf&X-Amz-Algorithm=AWS4
       HMACSHA256&XAmzDate=20190607T091144Z&XAmzSignedHeaders=host&XAmzExpires=86400&XAmz
       Credential=1SXTY4DXG6BJ3G4DXHR2%2F20190607%2Feuwest1%2Fs3%2Faws4request&XAmzSignature=
       a4645f58f16acf777416d72326c150992b7f739c53f604e23e7b447f34b00f28
[64]   P. online notes, "Mean value theorem," 2019. [Online]. Available:
       http://tutorial.math.lamar.edu/Classes/CalcI/MeanValueTheorem.aspx
[65]   Wikipedia, "Indeterminate form," 2018. [Online]. Available:
       https://en.wikipedia.org/wiki/Indeterminate form#Indeterminate form 0/0
[66]   C. A. D. Prado, "Tema 7, reglas de l'hospital," 2018. [Online]. Available:
       http://www.ugr.es/ camilo/calculo-ii-grado-en-matemat/apuntes/tema-7.pdf
[67]   Wikipedia, "L'hospital's rule," 2019. [Online]. Available:
       https://es.wikipedia.org/wiki/Regla de l%27H%C3%B4pital
[68]   ——, "Reciprocal rule," 2019. [Online]. Available:
       https://en.wikipedia.org/wiki/Reciprocal rule
[69]   J. Gorostizaga, "Universidad del pais vasco," 2019. [Online]. Available:
       http://www.ehu.eus/juancarlos.gorostizaga/apoyo/lim lhopital.htm
[70]   "Superprof material didactico," 2019. [Online]. Available:
       https://www.superprof.es/apuntes/escolar/matematicas/calculo/derivadas/derivacion-implicita.html
[71]   "Arkansas tech university," 2019. [Online]. Available: https://faculty.atu.edu/mfinan/2243/business33.pdf
[72]   Wikipedia, "Sequence," 2019. [Online]. Available: https://en.wikipedia.org/wiki/Sequence
[73]   S. Toida, "Basics of set," 2019. [Online]. Available: https://www.cs.odu.edu/ toida/nerzic/content/set/basics.html
[74]   Pr¥fWiki, "Definition: Sequence," 2019. [Online]. Available: https://proofwiki.org/wiki/Definition:
       Sequence#Formal Definition
[75]   Mathonline, "Bounded sequences of real numbers," 2019. [Online]. Available:
       http://mathonline.wikidot.com/bounded-sequences-of-real-numbers
[76]   Wikipedia, "Convergence," 2019. [Online]. Available:
       https://en.wikipedia.org/wiki/Sequence#Formal definition
[77]   H. Carrillo, "Sucesiones," 2019. [Online]. Available:
       http://www.dynamics.unam.edu/users/hcarrillo/archivos/Calculo1/Series%20y%20Sucesiones.pdf
[78]   P. Shunmugaraj, "Lecture 2 : Convergence of a sequence, monotone sequences," 2019. [Online]. Available:
       http://home.iitk.ac.in/ psraj/mth101/lecture notes/lecture2.pdf
[79]   Mathonline, "The boundedness of convergent sequences theorem," 2019. [Online]. Available:
       http://mathonline.wikidot.com/the-boundednessof-convergent-sequences-theorem
[80]   BRILLIANT, "Cauchy sequences," 2019. [Online]. Available: https://brilliant.org/wiki/cauchy-sequences/
[81]   F. A. Jr. and E. Mendelson, *CALCULUS*. SCHAUM'S outlines McGraw-Hill, USA.
[82]   Wikipedia, "Leibniz's test," 2019. [Online]. Available: https://es.wikipedia.org/wiki/Criterio de Leibniz
[83]   J. Belk, "Convergence of power series," 2009. [Online]. Available:
       faculty.bard.edu/belk/math142af09/ConvergencePowerSeries.pdf
[84]   N. Ltd, "Infinite series," 2004. [Online]. Available: http://www.nabla.hr/FUSeries1.htm
[85]   "Alternating series test," 2019. [Online]. Available:
       http://tutorial.math.lamar.edu/Classes/CalcII/AlternatingSeries.aspx
[86]   C. technical university in Prague, "Sequences of functions," 2006. [Online]. Available:
       http://math.feld.cvut.cz/mt/txte/3/txe3ea3a.htm
[87]   J. P. Delahaye, "The set of periodic points," *The American Mathematical Monthly*, vol. 88, no. 9, pp. 646–651,
       1981. [Online]. Available: https://www.jstor.org/stable/2320668?seq=1#page scan tab contents
[88]   A. Wade, "Math 401 - notes," 2005. [Online]. Available: http://www.personal.psu.edu/auw4/M401-notes1.pdf
[89]   P. Basarab-Horwath, "Uniform convergence," 2016. [Online]. Available:
       http://courses.mai.liu.se/GU/TATA57/Dokument/Uniform%20Convergence.pdf
[90]   Wikipedia, "Uniform convergence," 2019. [Online]. Available:
       https://en.wikipedia.org/wiki/Uniform convergence#To differentiability

[91]   A. A. Vian, "Sucesiones de funciones," 2019. [Online]. Available: http://www.ma.uva.es/ antonio/Teleco/Apun Mat2/Tema-7.pdf

[92]   J. O. S´anchez, "Sucesiones de funciones," 2019. [Online]. Available: https://www.cimat.mx/ jortega/MaterialDidactico/Analisis/Cap9v4.pdf

[93]   M. Lacruz, "Sucesiones de funciones," 2019. [Online]. Available: https://cafematematico.wordpress.com/2010/11/12/sucesiones-de-funciones/

[94]   R. S. of Arts & Sciences, "Calculus i integration: A very short summary," 2019. [Online]. Available: http://sites.math.rutgers.edu/~rainsfor/IntegralsWithInverseTrig.PDF

[95]   Wikipedia, "Antidifferentiation," 2019. [Online]. Available: https://en.wikipedia.org/wiki/Antiderivative

[96]   C. Polanco, *ADVANCED CALCULUS: FUNDAMENTALS OF MATHEMATICS*. Bentham Science Publishers, - Sharjah, UAE.

[97]   A. Akg¨ul, "Reproducing kernel hilbert space method based on reproducing kernel functions for investigating boundary layer flow of a powell–eyring nonnewtonian fluid," *Journal of Taibah University for Science*, vol. 13, no. 1, pp. 858–863, 2018.

[98]   ——, *Reproducing Kernel Method for Fractional Derivative with Non-local and Non-singular Kernel: Trends and Applications in Science and Engineering*, 01 2019, pp. 1–12.

[99]   ——, "A novel method for a fractional derivative with non-local and nonsingular kernel," *Chaos, Solitons & Fractals*, vol. 14, pp. 478–482, 2018.

[100]   ——, "On the solution of higher-order difference equations," *Mathematical Methods in the Applied Sciences*, vol. 40, no. 17, pp. 6165–6171, 2017. [Online]. Available: https://onlinelibrary.wiley.com/doi/abs/10.1002/mma.3870

[101]   ——, "A novel method for the solution of blasius equation in semi-infinite domains," *An International Journal of Optimization and Control: Theories & Applications (IJOCTA)*, vol. 7, p. 225, 07 2017.

[102]   ——, "A new method for approximate solutions of fractional order boundary value problems," *Neural, Parallel and Scientific Computations*, vol. 22, pp. 223–237, 01 2014.

[103]   A. Akg¨ul and E. Akg¨ul, "A novel method for solutions of fourth-order fractional boundary value problems," *Fractal and Fractional*, vol. 3, p. 33, 06 2019.

[104]   E. K. Akg¨ul, "Solutions of the linear and nonlinear differential equations within the generalized fractional derivatives," *Chaos: An Interdisciplinary Journal of Nonlinear Science*, vol. 29, no. 2, p. 023108, 2019. [Online]. Available: https://doi.org/10.1063/1.5084035

[105]   E. Karatas Akg¨ul, "Reproducing kernel hilbert space method for nonlinear boundary-value problems," *Mathematical Methods in the Applied Sciences*, vol. 41, no. 18, pp. 9142–9151, 2018. [Online]. Available: https://onlinelibrary.wiley.com/doi/abs/10.1002/mma.5102

[106]   B. Boutarfa, A. Akg¨ul, and M. Inc, "New approach for the fornberg–whitham type equations," *Journal of Computational and Applied Mathematics*, vol. 312, pp. 13 – 26, 2017, iCMCMST 2015. [Online]. Available: http://www.sciencedirect.com/science/article/pii/S0377042715004689

[107]   A. Atangana and A. Akg¨ul, "Can transfer function and bode diagram be obtained from sumudu transform," *Alexandria Engineering Journal*, 2020. [Online]. Available: http://www.sciencedirect.com/science/article/pii/S1110016819301760

[108]   A. Atangana, A. Akg¨ul, and K. M. Owolabi, "Analysis of fractal fractional differential equations," *Alexandria Engineering Journal*, 2020. [Online]. Available:http://www.sciencedirect.com/ science/article/pii/S1110016820300065

[109]   D. Varberg, E. Purcell, and S. Rigdon, *C´alculo Diferencial e Integral*. Pearson Educaci´on, M´exico.

[110]   StackExchange, "Limit (*an*) when $n \to ¥$," 2017. [Online]. Available: https://math.stackexchange.com/questions/882741/limit-of-1-x-nnwhen-n-tends-to-infinity

[111]   Wikipedia, "Geometric series," 2019. [Online]. Available: https://en.wikipedia.org/wiki/Geometric series

[112]   J. C. Verdes, "Series num´ericas de potencias," 2014. [Online]. Available: http://dm.udc.es/elearning/MaterialDocente/series con graficas.pdf

[113]   S. Exchange, "Why cos(*n*p) = (−1)*n*?" 2019. [Online]. Available: https://math.stackexchange.com/questions/911716/why-sinn-pi-0-andcosn-pi-1n

[114]   "Infinite series," 2019. [Online]. Available: https://resources.saylor.org/wwwresources/archived/site/wp content/uploads/2011/04/INFINITE-SERIES.pdf

SUBJECT INDEX

A

Absolute value 7
Absolute value inequalities 10
Algebraic functions 26,27
Algebraic maps 27
Alternating series 114
Antiderivative 125
Antidifferentiation 299
Arithmetic sequences 134

B

Basic antiderivatives equations 129
Basic limits of functions 54
Basic limits of maps 54
Bijective functions 24
Binomial series 111
Bolzano's theorem over functions 68
Bolzano's theorem over maps 68
Bolzano-Weierstrass's theorem over functions 68
Bolzano-Weierstrass's theorem over maps 68
Bounded sequences 98

C

Cauchy sequences 101
Classification of functions 22
Classification of maps 22
Closed intervals 5
Composition of functions 24
Composition of maps 24
Construction of irrational numbers 13
Continuity of a function 59
Continuity of a function using One-sided limits 61
Continuity of a function using Two-sided limits 62
Continuity of a map 63
Continuity of a map using One-sided limits 64
Continuity of a map using Two-sided limits 64

Convergent series 105
Cosine functions 34
Cosine maps 34
Critical points 83

D

Decreasing sequences 98
Definite integral 126
Dependent variable 16
Derivative of a function 73
Derivative of a map 79
Derivative of the reciprocal of a function 90
Derivative of the reciprocal of a map 90
Derivative test 83
Derivatives of main functions 77
Differentiability of a function 75
Differentiability of a map 81
Differentiation of implicit functions 86
Differentiation of sequence of functions 121
Displacement and distance traveled 134
Divergent sequences 99
Domain of a function 15
Domain of a map 18
Drug absorption 57

E

Elementary functions 26
Elementary maps 26
Existence of One-sided limits of a maps 47
Existence of Two-sided limits of a maps 48
Explicit functions 26
Exponential functions 36
Exponential function limit 53
Exponential map 44
Exponential map limit 44

F

Finite sequences 97
Finite series 97

Carlos Polanco
All rights reserved-© 2020 Bentham Science Publishers

www.ingramcontent.com/pod-product-compliance
Lightning Source LLC
Chambersburg PA
CBHW080020240326

41598CB00075B/522